Managing Security

Crime at Work

Volume III

Edited by

Martin Gill

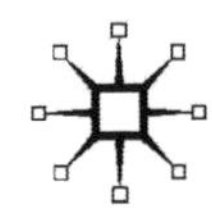

First published in 2003 by Perpetuity Press Ltd

Published by
PALGRAVE MACMILLAN

Palgrave Macmillan in the UK is an imprint of Macmillan Publishers Limited, registered in England, company number 785998, of Houndmills, Basingstoke, Hampshire RG21 6XS.

Palgrave Macmillan in the US is a division of St Martin's Press LLC, 175 Fifth Avenue, New York, NY 10010.

Palgrave Macmillan is the global academic imprint of the above companies and has companies and representatives throughout the world.

Palgrave® and Macmillan® are registered trademarks in the United States, the United Kingdom, Europe and other countries.

ISBN-13: 978–1–899287–65–9
ISBN-10: 1–899287–65–5

This book is printed on paper suitable for recycling and made from fully managed and sustained forest sources.

A catalogue record for this book is available from the British Library.

Printed and bound in Great Britain by
Antony Rowe Ltd, Chippenham and Eastbourne

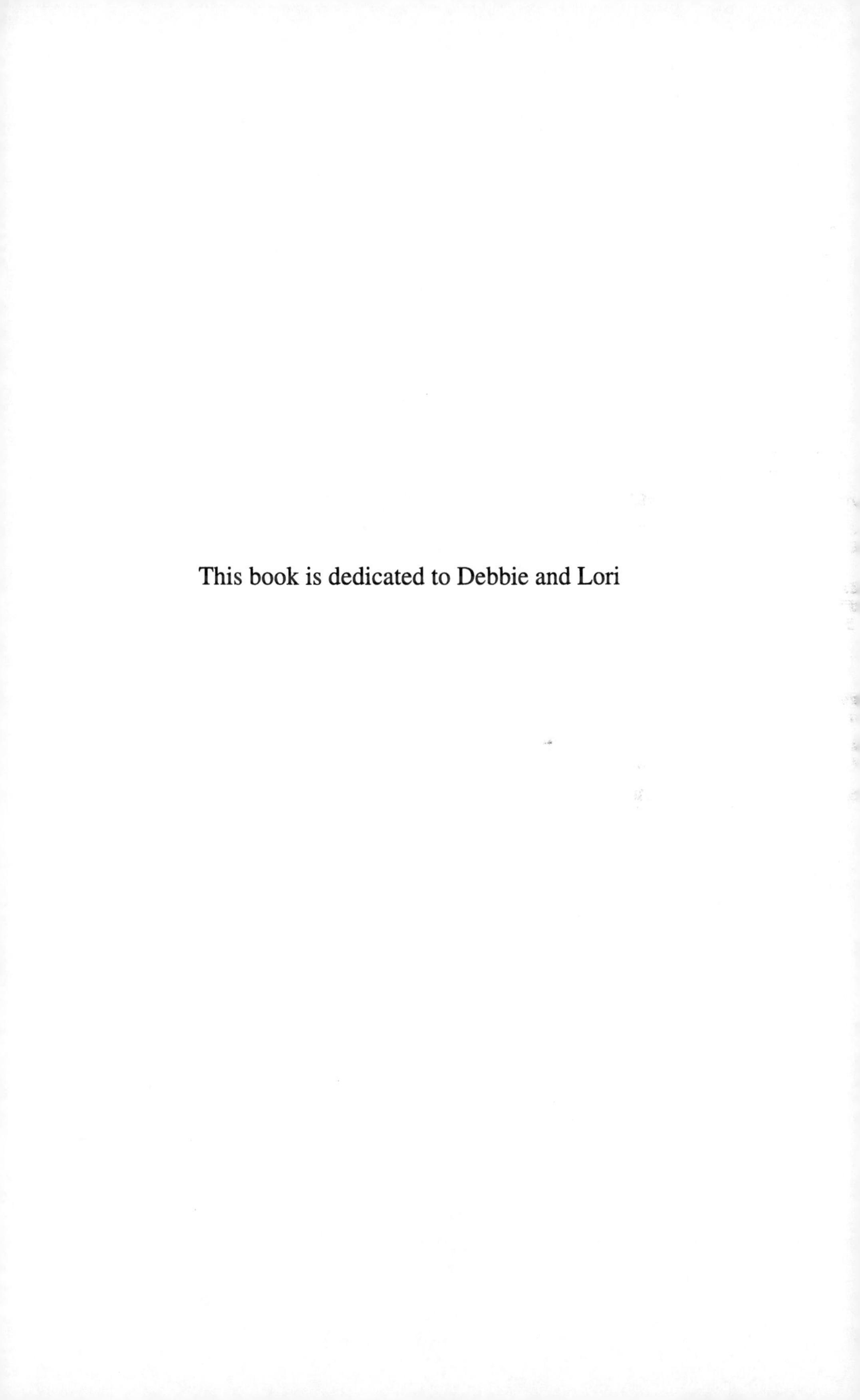

This book is dedicated to Debbie and Lori

Contents

Acknowledgements

About a year ago Adrian Beck suggested that it was 'about time you produced another *Crime at Work* book'. I had been thinking about it but his encouragement prompted me into action and I am grateful for this—Adrian appeared to have a point: I asked around and found that others felt there would be some interest too. In reality there are comparatively few able researchers interested in 'managing security', but it has been a source of great encouragement that many of those who are interested have agreed to contribute a chapter to this book. I am especially grateful to all the authors who responded to my emails with apparent enthusiasm. I remain exclusively responsible for any shortcomings.

I would like to thank the students of the Scarman Centre for continually showing so much interest over so many years. During the process of producing this book I changed roles: while I remain a Professor of Criminology at the University of Leicester I have set up a company jointly with the University—Perpetuity Research and Consultancy International (PRCI).

I would like to thank all those colleagues within the University who have encouraged and helped me. I would also like to thank the many practitioners and policy makers who in different ways contribute to managing security and who give up their time and money to support research; I hope that they in particular will find the papers of interest and use.

Finally, a special thanks to Karen, Emily, Karis and Oliver, who provide inspiration of the very best kind.

Martin Gill

Contributors

Joshua Bamfield is Director of the Centre for Retail Research, Nottingham, and was formerly Head of the School of Business at University College Northampton. Over the past 15 years he has carried out research into the costs of retail crime and electronic surveillance in Britain and overseas. He is responsible for the UK civil recovery programme whereby retailers obtain compensation from shop thieves.

Adrian Beck is a Senior Lecturer at the Scarman Centre, University of Leicester. His research interests include: crime against the retail sector, including violence against staff; evaluating security equipment; the use of CCTV; staff dishonesty and shop theft; and crime and policing issues in Eastern Europe. He recently completed the first pan-European study on the extent of the problem of shrinkage throughout the European 'fast moving consumer goods' sector, and is currently looking at developments in RFID tagging technologies.

Charlotte Bilby is a Lecturer in Forensic Psychology in the School of Psychology at the University of Leicester, and Tutor for courses in the assessment and treatment of sex offenders. Her research interests include the effectiveness of community penalties and offending behaviour treatment programmes, and the use of independent evaluations in governmental policy development.

Kate Bowers is a Research Fellow at the University of Liverpool. She has a broad interest in environmental criminology, and in particular quantitative methods for measuring criminological phenomena. Some of her recent published research has involved measuring geographical displacement and burglary reduction caused by crime prevention schemes. She is also developing new methods for quantifying the clustering of crimes in space and time, while retaining an interest in business crime, the subject of her doctorate.

Mark Button is a Senior Lecturer at the Institute of Criminal Justice Studies, University of Portsmouth. He has written extensively on private policing, publishing many articles and book chapters and completing two books with Bruce George: *Private Security* (Perpetuity Press) and *Understanding the Private Security Industry Act 2001* (Caltrop). As sole author he has also written *Private Policing* (Willan).

Paul Chapman, in the course of his doctoral work at the University of Warwick, focused on performance improvement in the automotive and aerospace industries businesses. Since joining Cranfield School of Management in 2000, as a Senior

Research Fellow, he has extended his experience into the 'fast moving consumer goods' sector. His research emphasis has a strong applied dimension, in particular through working with leading companies to identify how they can thrive in turbulent market conditions by employing the principles and practices of time-based competitiveness.

Louise Clare played a substantial part in the development of the funding application for the Business Crime Direct project. She continued to provide independent consultancy to the project until September 2001, and led the interim evaluation of the project up to that time. During her time as a Research Assistant in the Environmental Criminology Research Unit (ECRU), Department of Civic Design, University of Liverpool, she was also involved in the evaluation of the Burglary Reduction Initiative, and has a keen interest in the influence of urban design on crime.

Andrea Di Nicola has a doctorate in criminology from the University of Bari-Trento, and is now a researcher at Transcrime, the Research Centre on Transnational Crime at the University of Trento (Italy). He has a variety of research interests, including: migration and crime, and trafficking in human beings; economic and organised crime prevention; and the evaluation of crime reduction policies.

Nick Dodd is currently working as a business change consultant, having previously studied investigative psychology at the Universities of Surrey and Liverpool. He has taken a socio-cognitive approach to his research, and has highlighted a number of issues in the study of individual offenders, as well as the organisational environment which facilitates such criminal activity.

John Douglass is a Research Assistant in the Environmental Criminology Research Unit (ECRU), Department of Civic Design, University of Liverpool. He joined the Unit in June 2000, having completed the Master of Civic Design programme in the same department. Although his main research interests lie in the overlap between the fields of public health and crime prevention, he has provided independent consultancy to the Business Crime Direct project since June 2001, and is leading its evaluation.

John Gearson is a Senior Lecturer in the Defence Studies Department, King's College London. He is currently on secondment to the House of Commons Defence Select Committee as a policy adviser. Previously he worked as a management consultant, was a special adviser to the City of London Corporation on the terrorist threat to the City and advised the US Congress National Commission on Terrorism. He lectures on terrorism and defence policy, and is currently completing a book on terrorism for Polity Press.

Bruce George has been Labour Member of Parliament for Walsall South since February 1974, and a member of the Privy Council since December 2001. He has been Chairman of the House of Commons Defence Committee since July 1997. He has long campaigned for the regulation of the security industry, and since 1977 has introduced six versions of a Private Member's Bill on this topic. He has written a number of articles and books, including *Private Security* with Mark Button (Perpetuity Press).

Martin Gill is Managing Director of Perpetuity Research and Consultancy International (PRCI), and Professor of Criminology at the University of Leicester. His published work includes nine books, as well as over 50 journal and magazine articles, and he is co-editor of the *Security Journal*. He is currently undertaking research on robbers, thieves, fraud, the security industry, CCTV and false burglar alarms. He is a Fellow of the Security Institute, a member of the Risk and Security Management Forum, the Association of Certified Fraud Examiners, the British Society of Criminology, the Security Guild and ASIS International.

Read Hayes has been an independent security consultant with Loss Prevention Solutions Inc, in Winter Park, Florida, since 1984. He has over 25 years' hands-on experience in the control of crime and loss with numerous organizations worldwide. He is a Certified Protection Professional (CPP), Certified Security Trainer (CST) and Certified Fraud Examiner (CFE). As a Research Associate at the University of Florida and at the Loss Prevention Research Council, his research focus includes offender decision-making, supply chain protection and premises security.

Michael Levi has been Professor of Criminology at Cardiff University since 1991. He is a member of the Criminological and Scientific Council of the Council of Europe, for which he acts as scientific expert on organised crime. He has conducted many research studies for the Home Office and the private sector on business crime and its prevention and on money-laundering, organised crime and the proceeds of crime. His most recent book (with Andy Pithouse), *White-Collar Crime and its Victims*, will be published by Oxford University Press in 2003.

Jacqueline L. Schneider is a Lecturer in Crime Prevention and Investigation at the Scarman Centre, University of Leicester. For the past three years she has been working on crime reduction projects funded through the Home Office Crime Reduction Programme. Her latest project was with West Mercia Constabulary, serving as research manager for a Home Office-sponsored project aimed at reducing the stolen goods markets by implementing the 'market reduction approach'. Prior to her work in England, she taught undergraduate courses at universities in the United States, as well as working for a think tank. She has authored several papers and book chapters on gangs and stolen goods markets.

Martin Speed took a first degree in English from University College London, then worked in a variety of retail management roles before specialising in security and the management of information. In 1992 he obtained an MSc in Crime Risk Management from Cranfield School of Management. As Project Officer for the British Retail Consortium's retail crime initiative, he designed and managed the first Retail Crime Survey. He obtained his doctorate from Surrey University in 2001.

Geoff Taylor graduated with an MSc in Security Management & Information Technology from the Scarman Centre in 1993. He gained experience as a retail security manager and was instrumental in delivering a commercial burglary reduction strategy before working for Leicestershire police in a community liaison capacity. Since working on the money laundering study with Martin Gill, he has worked with the Home Office as a Senior Research Officer based in Nottingham. He now works as a Community Safety Co-ordinator in Birmingham.

Natalie Whatford is a Research Assistant to the Rt Hon Bruce George MP, specialising in areas of defence and security such as terrorism, and on issues relating to the Organisation for Security and Cooperation in Europe Parliamentary Assembly (OSCE PA), of which Bruce George is President. She has co-authored papers with Bruce George on responses to terrorism, and is also a member of the Royal United Services Institute, London.

Chris Young is a Research Associate in the Environmental Criminology Research Unit, Department of Civic Design, University of Liverpool, where he has worked for the last four years. He is involved in many aspects of the evaluation and understanding of crime reduction initiatives, and is experienced in the use of analytical tools such as GIS. Projects he has undertaken include interventions to reduce glass injuries resulting from assault—the Crystal Clear campaign—and 'alleygating', the latter aimed primarily at reducing domestic burglary via the rear of terraced properties. He has recently started work on assessing crime themes relating to New Deal for Communities areas.

Chapter 1

Introduction

Martin Gill[1]

This is the third volume of *Crime at Work*. The first two, published in 1994 and 1998, attracted significant attention, especially from the business community. Until these volumes appeared there had been few books containing empirical studies about understanding and preventing different offences committed at the workplace. While the main purpose of this section is to introduce the book, it begins with a brief review of the topic by considering events since 11th September, including developments on the prevention side and the potential implications of a regulated security sector.

Although workplace crime still remains an under-researched issue, studies in recent years have confirmed that the workplace suffers high levels of victimisation, and that for certain types of crimes, such as burglary, the workplace is proportionately more likely to be victimised than the home. Yet the finance sector and many businesses still lament the lack of interest shown by the police in the fraud cases they discover, and retailers still frown about the lack of police interest in attending the scene when a shop thief has been apprehended, or even when someone has been assaulted. Meanwhile, the police cite other priorities—those more at the forefront of public concern and on which they are judged. And we should not just focus on the police here: in the UK, Crime and Disorder Reduction Partnerships rarely include business and workplace crime within their remit, and this faulty prioritising—with isolated exceptions—seems common around the world.

Yet some things have changed. The tragic events of 11th September, etched in the minds of every living soul, have focused attention on security as an issue, not least the need for each corporation to protect all its assets and to develop contingency plans. It is no longer ridiculous to be concerned about planes being flown by terrorists into buildings.[2] This is a new era, where anything goes: suicide attacks where civilians are the victims are now real threats.[3] And other developments, not least in technology, have opened up new opportunities for crime: the ease with which money can be transferred around the world, for instance, makes things easier for money launderers and fraudsters.[4] A worst-case reading of the status quo would conclude that these new risks are looming large even while the traditional crime threats—including theft from shops, staff dishonesty, fraud and money-laundering, to name but a few—remain just as difficult to tackle, at least according to the evidence presented in this book.

So what has been happening on the prevention side? Well, quite a lot, in that new technologies offer new ways of tackling crime. The UK, almost alone in the world, has placed its faith in CCTV. One recent review has seriously questioned its effectiveness,[5] although too much scepticism is perhaps premature. Although very scholarly, the review took no account of the quality of the schemes, including the level of resources, quality of design, scale of 'buy in' of stakeholders, and effectiveness of management. These factors and others are crucial, and will hopefully be addressed by a more detailed study currently being undertaken.[6]

However, it is not just the effectiveness of technology that has been questioned. Security officers, according to some estimates, outnumber police officers, but where is the evidence that they are effective, or more cost-effective than the alternatives? The truth is that we still have very little evidence about what works. Perhaps one of the most incredible findings to emerge from contemporary studies of crime prevention is that so little attention has been paid to evaluation. Not only have there been few evaluations, but those that have been carried out have been deemed unreliable.[7] Thus we continue to implement projects and make the same mistakes, and we continue to put our faith in approaches which have not been properly tested.

In the UK a Security Industry Authority (SIA) has been formed, charged with regulating parts of the security world, starting with door supervisors and wheel clampers and then the contract guarding security sector (although oddly not in-house operations). This is an important development, and at the time of writing there is much optimism that its work will produce a credible alternative to the public police. In the past there have often been calls for some police tasks to be carried out by non-police personnel, but one of the main problems has been finding a credible alternative. Whether the SIA will emerge as the security world's 'fairy godmother' only time will tell. Certainly, contract guarding has long needed an invitation to join the 'the extended police family'. The SIA has the chance to set appropriate standards as a condition of granting a licence, but to bring about real change it will have to develop an inspection service with real teeth and convince sceptics inside and outside the industry that it can clear the marketplace of the cowboys (and for that matter, cowgirls).

The police have traditionally been sceptical of the private sector, although there is a noticeable change in attitude among those individual officers about to claim their pensions. Senior officers suddenly find that a new and well-paid career awaits them as corporate security managers, and others in lower ranks find the security industry offers a welcome and a rewarding living. I wrote a few years ago that the best credentials for a career in security are a pension from the police or military.

One might now add that an MSc and a professional qualification will also help, though they are not essential. Perhaps one of the best consequences or by-products of regulation will be a career structure for security. It is amazing to many that the security industry has never offered those leaving education a recognised career progression.

While regulation will have wide implications—a subject well beyond the scope of this discussion—there are at least two that merit comment here, as they relate to the attractiveness of the security function. The first concerns the implications of higher entry levels. Collectively, the security world engages in a wide array of tasks—as many as comparable public-sector agencies who appear far more able to attract good recruits. It seems inevitable that regulation will raise standards, and this could mean a need to engage a higher calibre of recruit; as a consequence the security industry could find itself recruiting from the same labour pool as other public-sector agencies, and would thus have to consider not only more attractive salaries but also other conditions of employment, of which a career structure is but one aspect. A second point is that while some security personnel undertake non-security tasks[8] the wider multi-skilling of operatives may help to sweeten the pill of increased costs that will inevitably follow from regulation. Whether this dilutes or enhances the security function is an important issue to watch.

There remains a dearth of research on security issues, particularly high-quality independent research. This is a serious limitation, and one that needs to be addressed. The *Security Journal* continues to publish some cutting-edge papers from around the world, but although such material is now more accessible,[9] it is still not abundant. This has implications way beyond the whims of academia: it means that there is no developed body of knowledge to guide practice, frameworks have not been systematically evaluated, new ideas have not been developed, and so on. There is insufficient research to inform and frame ideas, concepts and theories, a gap that this book seeks in some small way to help fill.

The book

This book contains 13 chapters; they all report new findings and new ideas (or sometimes older ones reframed), and they all seek to develop our understanding of how a range of security issues can be better managed. The papers have been selected because in different ways they contribute to understanding aspects of 'managing security' in the modern world, with all its new challenges. These are all scholarly papers which generate some important practical lessons. The early chapters consider management issues; later ones move on to consider specific strategies, where technology emerges as a key means of tackling different types of crime. Chapters 10 and 11 tackle different types of methodological issues, while the last two discuss the threat posed by new terrorism.

In the first chapter John Douglass, Kate Bowers, Chris Young and Louise Clare discuss 'Business Crime Direct', a scheme which aims to reduce crime against small and medium-sized businesses. Its experience offers important implementation lessons for those charged with tackling business crime. It also illustrates that it remains difficult to reach many businesses who have bad previous experiences. Here the involvement of credible partners such as the Chambers of Commerce has been found to be invaluable.

The second chapter, by Read Hayes, also looks at co-ordination issues, focusing on the views of senior loss prevention managers in the US. Worryingly, but perhaps not surprisingly, Hayes finds that major decisions are not supported by a recognised body of knowledge: managers place more emphasis on anecdote than on collecting and analysing incident data, formulating evidence-based strategies and measuring actual performance and impact. While one of these managers' major concerns for the future is staff dishonesty, the true road ahead may rest more in Hayes's observation that loss prevention managers are increasingly being asked to co-ordinate non-security functions such as audit, safety, building maintenance and even human resources.

In the third chapter Adrian Beck considers the viability of a much-heralded system to help management tackle shrinkage: automatic product identification (Auto ID). Beck's analysis suggests that it has greater potential for tackling non-malicious forms of shrinkage, such as process failures, than the threats posed by staff and customer theft. Nevertheless, Beck welcomes its overall contribution, not least its ability to track every product throughout the supply chain, facilitating the formation of an information-led strategic approach to shrinkage.

The use of technologies in tackling payment card frauds forms part of the discussion in the next two chapters. In Chapter 4 Mike Levi notes the changing patterns of this type of fraud and the complications that arise for prevention as a consequence. One important observation that Levi raises—frequently forgotten in studies of crime prevention, because the commercial imperative is rarely considered—is that some technological solutions are likely to be resisted by retailers. At present retailers do not lose much from fraud, but card companies do, and while chip cards with a PIN might help card companies, retailers can see transactions taking longer and customers being served less quickly, important issues in their business.

In Chapter 5 Geoff Taylor and I report on a study of money-laundering. Money-laundering has for some time been a priority for action across the world, although the events of 11th September certainly accelerated the introduction of new initiatives, especially in the US. The chapter considers finance companies' experiences of defining a 'suspicious activity', and the role of information technology in helping to identify suspicious transactions. The use of IT systems is attracting considerable attention as a potential means of differentiating the (presumably few) illegal transactions from the billions of perfectly legal ones. However, the findings revealed some considerable scepticism about the effectiveness of such systems, especially their cost-effectiveness, and some naive optimism from the inexperienced. In looking at the reasons we suggest caution in the adoption of an over-critical stance towards technology, when a lack of clarity about what is suspicious remains so pervasive.

In Chapter 6 Andrea Di Nicola discusses a very different type of crime at work: trafficking in human beings, and specifically in women for the purposes of sexual exploitation. Di Nicola bases his analysis on documented cases of the trafficking of women from Eastern European countries to Italy. Of particularly relevance to this book is his finding that such trafficking is facilitated by 'legitimate' actors, including taxi drivers, hotel owners, and travel and employment agencies. Indeed, Di Nicola argues that prevention strategies need to include a closer regulation of their activities.

Chapter 7 continues with a focus on Europe, where Joshua Bamfield reports on the first pan-European retail crime survey. He found that the rate of shrinkage varied between 0.87 and 1.76 per cent of turnover, that large retailers spent €5781 million on security and loss prevention—equivalent to almost one quarter of their total losses from crime—and that the total cost to retailers of theft and other crime was €29,599 million, equivalent to €76.83 a year paid by every person in the countries surveyed. Retailers regarded 82.4 per cent of shrinkage as being caused by *crime-related* behaviours, the two largest being theft by customers (45.9 per cent of shrinkage, or €13,210 million) and theft by employees (28.5 per cent, or €8238 million). These two types of theft are discussed in the next three chapters.

In Chapter 8 Jacqueline Schneider discusses shop thefts committed by prolific burglars, and found this to be common, especially to obtain property needed for stolen goods markets but not available through burglary. Schneider argues that these thieves are rational, in the sense that they know what they intend to steal prior to entering the shop. She notes that shoplifting is a major contributor to the stolen goods markets, and that if stores cannot prevent thefts then another option rests with the control of those markets.

Chapters 9 and 10 are about staff dishonesty. In Chapter 9 Martin Speed's work is based on an assessment of company records of detected employee offenders, and on a survey of staff to elicit their views. Speed found young and new employees to be more frequently caught, but that longer-serving employees perpetrate more costly and sophisticated offences and are the most confident of avoiding detection if they behave dishonestly. Speed develops a four-way classification of employees, and suggests a management strategy for tackling dishonesty in each group.

Nick Dodd develops this work in Chapter 10, and he too reports that there are different types of offender who merit different responses. His work is based on the analysis of files retained by companies on their employee offenders. As Dodd explains, this approach is controversial, but his conclusion is that, despite some limitations, they are a rich source of information, based as they are on actual cases rather than on opinions; and he explains his methodology for making the most of them. His findings offer important insights into managing staff theft.

The development of a methodology, this time a 'process-orientated' approach to tackling shrinkage, is the subject of Chapter 11. Here Adrian Beck, Charlotte Bilby and Paul Chapman discuss a prescriptive series of seven steps that form a stock loss reduction 'roadmap'. The guide acts like a manual, describing the overall activities that need to be undertaken in order to reduce stock loss. It consists of a general approach made up of the steps a company needs to follow, together with techniques and tools to help undertake each phase and to deal with problems that may be encountered. The structure is systematic, and provides a means of planning and undertaking stock loss reduction projects while guiding users towards continuous improvement through the cycle.

The last two chapters focus on terrorism, and in particular on lessons for the commercial world in dealing with different types of terrorism. In Chapter 12 John Gearson analyses the Provisional IRA's attempt to disrupt the financial sector, and the latter's response. What emerges is a scenario where too many firms were found to have no disaster or contingency plans prepared, and there was something of a cultural rift between public- and private-sector agencies. Gearson suggests a number of remedies based around co-ordinating a proper understanding of the threat amongst different groups and a shared commitment to post-event plans.

In the final chapter Bruce George, Mark Button and Natalie Whatford develop the theme of 'new terrorism' following 11th September, and they note the very real threat that terrorists will carry out atrocities resulting in mass casualties. The main target is likely to be businesses, and given the need to provide a safe working environment they could be liable if they do not put in place appropriate prevention strategies. The chapter outlines some of the weaknesses in current practice and focuses on the more significant areas for action. Their suggestions must be heeded, especially as complacency abounds, and the consequences for not doing so are too awful to contemplate.

One thing that should not be forgotten in any discussion of 'crime at work' is that sometimes organisations and its employees are the offenders. There is now a wealth of studies that document this,[10] and much more needs to be done—though that is beyond the focus of this volume, whose aim is rather to provide new insights into how crime threats to organisations can be better managed. Indeed, running through the papers in this book is the recognition that the process of 'managing security' is dependent on having good management information (intelligence, data and research). Security strategies of the future need to be based on new insights, good data and quality research across the breadth of security topics. In this book all the papers contribute in different ways to facilitating just this.

Notes

1 Thanks are due to Adrian Beck, Tony Burns-Howell and Martin Hemming for very helpful comments made on earlier drafts of this Introduction.

2 Although this risk is often already managed by companies which are located on flight paths.

3 See chapter 13.

4 See Chapter 5.

5 Welsh, B.C. and Farrington, D.P. (2002) *Crime Prevention Effects of Closed Circuit Television: A Systematic Review* . Home Office Research Study No. 252. London: Home Office. See also Gill, M. (ed.) (2003) *CCTV*. Leicester: Perpetuity Press.

6 A major evaluation is being undertaken by a team of researchers led by the author.

7 See, Tilley, N. (ed.) (2002) Analysis for Crime Prevention. Vol. 13, *Crime Prevention Studies*. New York: Criminal Justice Press.

8 See Chapter 2.

9 The 'security and risk database' provides instant access to security-related abstracts from papers in peer-reviewed journals (see *www.perpetuitygroup.com*).

10 For a recent example, see Smith, R. (ed.) (2002) *Crime in the Professions*. Aldershot: Ashgate.

Chapter 2

If You Don't Call Us, We'll Call You: The Experiences of Business Crime Direct

John P. Douglass, Kate J. Bowers, Chris Young and Louise Clare[1]

Business Crime Direct is an initiative funded by the Home Office's Targeted Policing Programme which aims to reduce crime against small businesses throughout Merseyside. It was envisaged that this initiative would build on the experience of previous research, which had found that it was crucial to access those harder-to-reach businesses that saw crime as an inevitable part of everyday life; these were often the most victimised businesses. It was also seen as important to provide a continued support service to those at the greatest risk, to maintain and extend the longevity of the effect of any assistance that was given and to increase the momentum of crime prevention efforts. BCD was therefore set up using a call centre to manage potential recipients of security improvements and to grant assistance. It would do this by receiving inward enquiries and making outward contact with businesses, and referring those of the latter most at risk to Security Advice Officers, who then made site visits. This dual approach to managing the contact between businesses and crime prevention experts, along with accurate recording of the history of this contact, should assist in providing such long-term support to hard-to-reach higher-risk businesses. The extent to which the service achieved these aims, and the problems experienced in implementing such an innovative scheme, are discussed in this chapter.

Context

Crimes against businesses have traditionally been afforded less priority than those against individuals and domestic properties, in terms of the attention given to them in academic research and national policy, and therefore also in terms of the number and scale of initiatives targeted towards them. This situation is now changing. In line with the Labour government's pro-business stance, there has been a realisation that economic development is strongly dependent on the attractiveness of an area for inward investment of large businesses. The prevalence of crime plays an important part in this. Furthermore, the additional costs posed to small businesses in high-crime areas, costs which include the need for premises security and inflated insurance premiums, impact strongly upon the business's viability. In order to survive, costs are ultimately passed on to the consumer. Yet in the many cases where the closure of businesses becomes an unavoidable reality, there is a strong consequential effect on the physical, social and economic environment, which often contributes to a spiral of decline.[2] Reducing the attractiveness of an area to new and relocating businesses (and residents) undermines regeneration efforts, and the accessibility of vital services to local residents is also reduced. The problem of business crime is now rising up the policy agenda, having been championed by the British Chambers of Commerce (BCC). The Home Office has also acknowledged the need to tackle business crime, and has responded with the production of a Business and Retail Crime Reduction Toolkit.[3]

Nevertheless, the existence of timely and accurate information remains problematic. Prior to the Chambers' survey, the 1994 Commercial Victimisation Survey,[4] supplemented by annual updates to the British Retail Consortium's survey, represented the best information available.

This comparative lack of information presented problems for the efficient targeting of resources on business crime in Merseyside. Following two phases of its Small Business Strategy, the Safer Merseyside Partnership (SMP) therefore sponsored a major three-year piece of research, conducted by Bowers at the Environmental Criminology Research Unit (ECRU) in the University of Liverpool.[5] The research, which included over 1500 risk assessment and victimisation surveys, produced some staggering findings, including the following: a hard core of business premises experienced most of all non-domestic crimes—21 per cent of all non-domestic burglaries were repeats, and 34 per cent of the victims had been burgled three or more times in the previous year. The findings led to obvious operational recommendations: a substantial impact could be made upon business crime by identifying and improving the security of a very small number of businesses. Hence the important need for reliable data in managing the crime prevention process, by directing security improvements to those most in need, was reaffirmed.

These findings created the rationale for the development of the Business Crime Direct (BCD) scheme. With grant aid having been distributed to businesses through the SMP Small Business Strategy, the research also informed the design and approach to some key elements of the service. For example, it introduced the idea of reducing all potential targets of commercial crime to a manageable sample by employing sifting criteria. Similarly, it observed the phenomenon of 'grant chasing', and emphasised the need to reach out to the most vulnerable businesses that do not have the ability to prevent crimes against themselves, and in many cases do not see the benefits of attempting to do so. It saw the overcoming of this 'learnt helplessness' as central to the reduction of crime against businesses.

Vision and objectives

The vision of the BCD service was an ambitious and well-considered one, supported by considerable research. It sought to implement technological, innovative and problem-orientated solutions to reduce crime against non-residential properties by the establishment of a Small Business Security Advice Service. Alongside its more tangible aim of delivering crime prevention advice, and measures to alleviate some of the pressure placed on police Crime Prevention Officers, the service also sought to raise the small business sector's awareness of the effectiveness of implementing crime prevention measures. In doing so, it attempted to combat 'learnt helplessness'. The objectives by which it proposed to achieve this vision most notably included:

- the introduction of innovative partnership structures, communication systems and revised working practices. Most notably, BCD includes Chambers of Commerce and Industry on Merseyside—the scheme is hosted by Liverpool Chamber of Commerce and Industry (LCCI)—and a pre-existing Intermediate Labour Market (ILM) training scheme.[6]

- the use of an (ILM) call centre to undertake 'Crime Risk Assessments' in the diagnosis and prioritisation of vulnerable small enterprises, supported by on-site surveys carried out by security risk assessment officers. The service would supplement this diagnostic role with continued support to vulnerable businesses through the maintenance of a customer care strategy.

The proposal for the service was particularly innovative, for several reasons. First, it employed a sophisticated 'sifting' process to prioritise those businesses most in need of assistance, thereby ensuring that public funding was targeted most efficiently. In addition to the advice offered during this process, it made available grants for security improvements to those most in need. Second, it proposed that ongoing care be offered to the most vulnerable of the businesses that had become members of the service, in order to support and nurture them up to a level of self-sufficiency in business crime prevention. Third, it harnessed new technology to increase the service's accessibility to and usability by businesses. By using a call centre, the scheme was able to offer a level of assistance, tailored to the needs of each business, at reasonable cost to the service. It therefore met all the criteria of the government's 'e-government agenda'. Finally, the scheme built new bridges between the public and private sectors via the inclusion of a range of partners, including the police, a Single Regeneration Budget Partnership, and most notably the Chambers of Commerce, for whom business crime is an increasingly important issue.

The application for Home Office funding of the BCD acknowledged the highly experimental nature of the initiative, and therefore placed emphasis on the need to 'learn from experience'. Furthermore, it stressed that the elements of the service that 'work well' would be identified by monitoring it closely.

The intended operation of the service

Figure 1 illustrates the cycle through which a business engaged by the BCD service would progress from initial contact to the completion of a full risk assessment.

The complete cycle through which a business interacts with the service can also be described as a multi-stage process, involving: the engagement of the business; the diagnosis of vulnerability; risk assessment and advice; and follow-up support.

Figure 1. BCD's risk diagnosis cycle

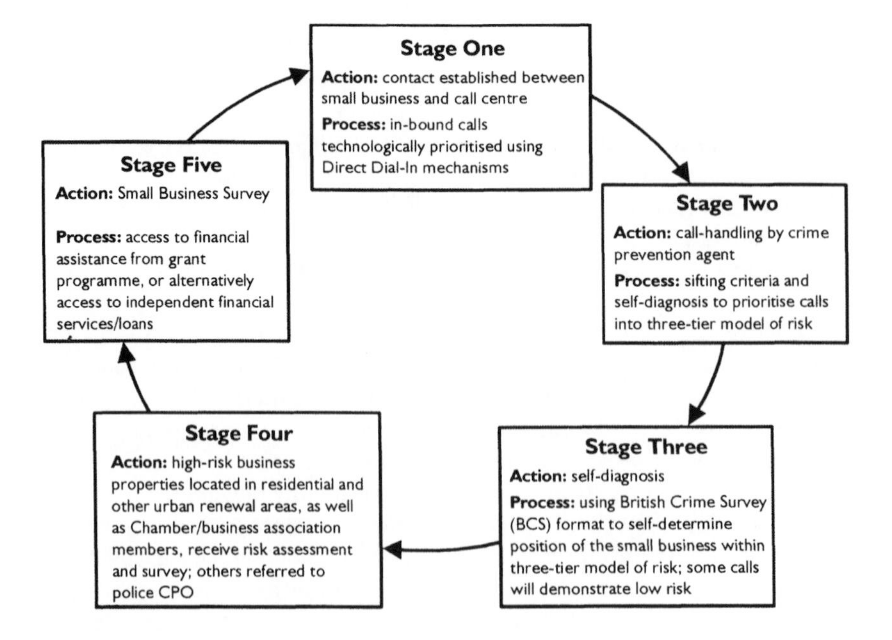

Source: adapted from Small Business Security Advice Service On-line Partnership (2000) *Targeted Policing Development Plan.* Unpublished application submitted to the Home Office.

Engaging the business

It was envisaged that most users of the scheme would be engaged by the service through its call centre. The primary role of the call centre would be to receive incoming calls from users who had been made aware of the service through its marketing, all of which featured the freephone number. Additionally, the call centre's role would include outbound calling to businesses referred to the service by other means. The main source would be the police recorded crime system, and businesses that had experienced repeat victimisation would be prioritised.

Diagnosis of vulnerability

Upon answering a call, the call centre would ask a series of questions of the caller, from a predetermined script. The answers to the questions were recorded, and automatically generated a 'vulnerability score', which in turn determined the 'level of service' to be received by the business: 'Gold', 'Silver' or 'Bronze' for high-, medium- or low-risk businesses respectively. It was intended that Gold businesses would receive the highest priority attention, access to the most services, and continued support to bring them up to minimum risk levels.

The factors upon which the diagnosis was based included the business's crime history, with greater emphasis placed on recent burglaries and repeat victimisation, and its location, ie whether this was within highly deprived areas or crime hotspots. Additionally, businesses were prioritised on the basis of whether they were members of a business association.

It was envisaged that those businesses awarded Gold or Silver service would be reviewed after three years, after which time it was hoped that the crime prevention assistance that they had received would have reduced their vulnerability sufficiently for their demotion to Bronze status.

Risk assessment and advice

Having diagnosed the vulnerability of a business, appropriate action would be taken by the call centre to offer it a service. If the business was found to be 'appropriately' vulnerable (Gold or Silver), an appointment for a visit from a security advice officer (SAO) would be made, and the details of the business would be referred to the risk assessment service.

The two SAOs employed by the service would receive referrals from the call centre, and visit the premises to undertake a full risk assessment. The visit would provide an additional opportunity to collect data, which could later be used for research purposes, about those factors influencing a business's vulnerability to crime. Following the visit, a report would be issued to the business, which would then be invited to submit to the service three quotes, from security suppliers included on a register (to be developed by the service) of approved security products and services. The initial vulnerability criteria, combined with the additional factors investigated during the on-site survey report, would then be used as a basis for deciding the allocation of a maximum £1000 grant. Irrespective of their decision to supply three quotes, users of the service were therefore given a full risk assessment, documented in report form.

Follow-up support

The final intended component of the service was the ongoing support programme, envisaged as offering a similar kind of service to that enjoyed by a member of a car breakdown association. In addition to risk assessment, information on the possible allocation of a grant, and the availability of security advice over the telephone, this service would include regular newsletters or reports, detailing current trends in crime against business and appropriate ways of addressing them.

The implementation process

The operational phase of the scheme began in January 2001 with the appointment of two joint general managers. The first surveys were conducted in April 2001, following an official launch in March. The following paragraphs outline the experience of implementing the scheme. For ease of analysis, this experience has been presented under several issue-based headings.

Engagement to the service

The main mechanism by which users were to be engaged to the service was via the call centre hotline. This had two main tasks: the receipt and processing of incoming calls, and outbound calling.

Incoming calls. The idea of using a call centre was based on current trends in the customer service industry. Importantly, the 'self-selection' approach represented the only really viable solution, given the external limitations placed on the project (these are explored later within the 'outbound calling' section of this chapter).

Calls to the hotline were to be generated by marketing and publicity of the scheme and its freephone number. The marketing channels adopted can be seen in Figure 2.

The marketing campaign was made more difficult because its 'universe'[7] was unknown. This may help to explain why it was not especially successful. The hotline generated an average of nine calls per week between April and July

2001 (a rate of approximately 470 per year). At the funding levels originally set out in the delivery plan, the 'per call' cost of the call centre service was therefore considered to be too high.

The effectiveness of marketing is notoriously difficult to monitor, and cost benefit analysis is even more difficult, particularly in public-sector schemes. In order to monitor the effectiveness of the marketing, the question 'where did you hear of the service?' was asked of every caller to the hotline. Unfortunately this data remains difficult to extract, and is not currently available. Even so, the information would not explain why so few people rang the call centre, but merely the relative effectiveness of each channel within the population that had responded. The effectiveness of the various marketing approaches therefore remains unclear.

Figure 2. Marketing BCD

- A well-attended launch at Liverpool Football Club, featuring agencies involved and local businesses.
- A poster campaign on buses and in bus shelters.
- Press releases, including several spreads in the business sections of local newspapers, inclusion in the *Financial Times*, and profiles within the security-related press.
- Broadcasts on Radio Merseyside.
- The distribution of promotional material, including posters, coasters, stationery and folders.
- A glossy brochure featuring crime prevention advice (self-funded via the inclusion of advertising).
- Cards distributed via Crime Prevention Officers.
- Promotional material placed within Costco as 'business of the month', and cards distributed with Costco application forms.
- The development of a website[8] (which enables on-line appointment booking).

There is insufficient evidence to state whether the lack of response to the marketing campaign was due to theory or implementation failure. Both were suggested by members of the steering group. The proponents of 'theory failure' argued that people simply do not like to use call centres, whereas those who believed that the limited response was due to 'implementation failure' claimed that the marketing may not have been sufficient or correctly targeted, and suggested that the use of different marketing channels should be pursued within the existing strategy.

Outbound calling. The effective delivery of referrals to the call centre for outbound calling was hindered by the limitations of Merseyside Police's crime recording system.[9] These are largely symptomatic of the lower priority traditionally given to the prevention of business crime, and are replicated in many police forces across the UK.[10] Closer scrutiny of the system led to the discovery that it has no dedicated field to indicate that a crime has occurred against a business. Despite the substantial progress made by 'Operation Goodkey', a training initiative which sought to encourage the marking of a 'C' for 'commercial crime' within the gender field of a crime report form, identifying a business crime remains problematic.

The service was eventually able to receive referrals from the police system following the signing of a data-sharing agreement with the police. Crimes recorded as 'burglary other' (which include crimes such as theft from sheds) are downloaded daily and eligible businesses are extracted manually. Each of these businesses is then contacted by one of the BCD's SAOs to ensure its compliance, and visited for a risk assessment. Despite its usefulness in identifying recent victims of crime, there are several problems with this approach. First, the reliability of the downloads has never been rigorously tested, and the effect of local differences in crime-recording practices remains unclear. Second, the approach generates referrals only for those businesses that have been burgled. Nevertheless, it satisfies the need to deal quickly and efficiently with businesses which have been shown to be susceptible to repeat burglary, due to their recent experience of an incident.[11] It therefore identifies the most vulnerable businesses in respect of one causal factor in that vulnerability.

A further problem that faced the service in terms of engaging users to the service lay (and continues to lie) in its inability to pinpoint exactly the nature and location of its potential users: businesses. This is by no fault of its own: there remains no comprehensive and up-to-date register of businesses on Merseyside. Despite useful ongoing work to develop a National Land and Property Gazetteer, the sources currently available are uncoordinated and do not offer full coverage of Merseyside. Even the most potentially useful sources for such information are exclusive: VAT registers held by Local Authorities store information relating only to businesses with an annual turnover greater than £55,000.

In this respect, the service's relationship with the Chambers of Commerce proved invaluable. LCCI's database contains details of almost 30,000 businesses across Merseyside. In order to stimulate activity, outbound calls were made to 429 of the most vulnerable businesses on LCCI's database—those which were members of

Merseyside's six business associations, set up specifically to support businesses in the most deprived areas[12] and supported by Merseyside Chambers. However, this generated a response from only 48 firms (an 11 per cent response rate): 21 requested literature, and 27 made appointments. The hosting of the scheme by LCCI is also thought to be advantageous in respect of the perceived legitimacy of the service amongst businesses. Use of the LCCI name in outbound calling helped to convince many suspicious businesses that the call was not another sales pitch from a private security firm.

Management/organisational culture

Within all multi-agency partnerships, conflicts and tensions are caused by the differing objectives of the partnership's participants, who aim to divert the project in various different directions. Research institutions for example, are commonly criticised for requesting data as an output of projects, for analysis and research purposes, whereas local authority officers may be more likely to steer projects in directions which gain political support for the councillors they serve.

The multi-agency BCD partnership (the partners in this scheme are collectively referred to in the remainder of this text as 'the steering group') is no different. Examples of the need to meet wider objectives include: the ECRU's desire to further research on business crime; the need for Southport Telematics to fulfil the objectives of their ILM training initiative; and the desire of the Chambers of Commerce/business associations to increase satisfaction amongst their members by providing an effective response to the growing problem of business crime.

Tensions arose during the pilot phase when it became apparent that the strategy was not leading to the levels of demand envisaged. Partners were quick to diagnose the cause of the problem as lying with the element of the service for which they were individually not responsible, and communication between elements of the service worsened. Particular problems arose from the lack of clear definitions of the role of each partner. The management of the hotline, for example, were initially reluctant to return the funding out of which they had staffed the call centre simply because of a lack of incoming calls, when their role had been defined as 'the provision of call centre services'. Having no involvement with the promotion of the hotline number, they could hardly be blamed for the telephone failing to ring.

The issue of accountability appeared further to frustrate partners who all had differing views on how the scheme should be implemented. Accountability for the scheme was ultimately to the steering group, which was not always fully attended and at times could not agree on successful ways forward. This applied not only to the general direction of the scheme, but also to issues which were thought more trivial, such as that of data collection.

Finally, through the proposal of different options, and their evaluation by the steering group, the strategy was amended and amicable solutions found. If nothing else, the experience led to a greater understanding of the need to conduct an exhaustive risk analysis and contingency planning exercise to highlight both positive and negative scenarios.

Data collection and use of evidence

The well-documented problems[13] of identifying and then monitoring business crime, which cannot be separated from its traditionally low political priority, are undoubtedly contributory factors in the evident dearth of research on business crime. This is one of the reasons that the BCD service set itself ambitious objectives in relation to data collection. The other reason was that the service intended to use this data to perform 'real time' calculations of a business's vulnerability on the basis of identified trends,[14] the results of which would generate a graded and prioritised response. Thus the importance of efficient data collection and processing was paramount. However, the reality of the data collection process differed from that envisaged for a variety of reasons:

- The original script for collecting data was deemed too long by the steering group, and several data items were removed.

- The ILM staff were not comfortable asking many of the questions, due to the latter's sensitivity and to the emotional state of many recently-victimised callers. This highlights a training issue for staff, and the need accurately to communicate the purpose of the service to users, some of whom demanded an emergency response.

- The data collection process was jeopardised by an initial removal of the need to prioritise businesses using vulnerability criteria. This was due to a number of factors: first, the low demand initially expressed meant that every eligible business was visited; second, members of the steering group were uncomfortable with labelling businesses as only Bronze or Silver members, and the grading of service was abandoned. Although the allocation of grant would still vary by need, all eligible recipients would experience a similar response and level of service from BCD.

The low overall demand for the call centre and the initial removal of the need to prioritise were perhaps the main reasons for the downfall of the scheme as originally intended. The perceived role of the scheme was altered as to both its management and its users, from the BCD 'club' with a commitment to a sustainable customer service strategy to that of a 'one-off' security improvement scheme. The removal of the need for the sifting process also meant that the timely inputting and processing of data was not required. Consequently, the collection of some items of data was seen to place an unnecessary burden on the time of the call centre staff and the SAOs. Additionally, the adoption of a 'one-off' approach removed the need to recall customer details and the history of contact with the service. Thus the urgent need for an efficient processing and customer management database at the call centre was also reduced.

Operational problems also impacted upon the data collection process. The need for sophisticated, compatible and linked databases at the two physically remote sites was not considered in sufficient detail and sufficiently in advance of the beginning of the scheme. The availability of skilled personnel was also a problem. The result was that the call centre element and the central BCD operation failed to work effectively together to enable efficient transfer of information. The lack of a substantial two-way information link between the two elements of the service also meant that basic intended functions of the call centre, such as the booking of appointments, could not be implemented effectively.

Options evaluation and service evolution

The low demand for the call centre, and the associated frustrations that this caused, led to the need for an options evaluation and appraisal exercise. In July 2001 it was decided that partners could present their options for remedying the situation. A discussion paper dealing specifically with the issue of redistributing resources away from the call centre was presented by LCCI, and a response and a wider discussion paper, in which the options proposed by all other partners were collated, was also presented by the University.

In addition to the low take-up of the service, several other factors contributed to the change of strategy. BCD management had found that networking with business associations and local area partnerships had produced a great number of referrals. Although these were not subject to the initial vulnerability diagnosis, their eligibility for funding was explored prior to the issuing of a grant. This approach had led to the formation of a working relationship with the Netherley Valley SRB partnership, whose funding had enabled the appointment of a dedicated SAO for the partnership area. Demand for group schemes had also been expressed, for example by the Chinatown area, which wished to group together the maximum funding available to it from BCD to contribute to an area-wide CCTV scheme. Similar requests for assistance had also been expressed by an area of St Helens, and BCD had repeatedly identified the merit associated with businesses 'tackling crime together' during discussions with businesses whilst conducting risk assessment surveys.

Trends within crime pattern analysis—increasingly influencing operational policing—had also placed emphasis on the need to acknowledge and respond to spatial concentrations of crime, or 'hotspots'; this need is continuing.[15] Indeed, research in Merseyside with which BCD was involved had even identified 'hot ribbons' of crime occurrence along arterial routes.

Thus the need for a strategy shift, highlighted by the need to redistribute resources from the underused hotline, presented an opportunity to concentrate the efforts of BCD officers on setting up 'business clusters' and facilitating the implementation of group schemes such as CCTV and Radiolink systems. The change of strategy was approved by the fundholder, SMP, in August 2001.

The move to reducing the vulnerability of business clusters has some distinct advantages. This is because businesses that have been target-hardened in isolation from those around them are unlikely greatly to benefit other businesses in the vicinity. However, when fairly intense measures are implemented in a smaller area, there is often a diffusion of benefit such that the vulnerability of those close by but not directly involved in the scheme also decreases.[16]

Despite the shift, the core activities of the service as originally intended have been maintained. Referrals continue to come from the police system, and the freephone number, which has now been redirected into the BCD office, is answered by core BCD staff.

Satisfaction with/Effectiveness of the service

At the time of writing, a full quantitative evaluation of the service has not yet been conducted; it is due to be delivered in early 2003. It is intended that this will include an evaluation questionnaire indicating victimisation levels since intervention by the service. Focus groups, with a sample of the cluster beneficiaries, will also be used to assess the usefulness of the interventions. Despite its limitations, attempts will also be made to utilise recorded crime data to identify repeat incidents affecting beneficiaries of the service.

The 'continued customer support' objective set out within the application for funding has yet to be met. The main vehicle for this was originally intended to be the call centre, as 'members' of the service (Gold, Silver and Bronze) were able to access the hotline to gain crime prevention advice and more detailed risk assessment. However, this approach was undermined by a combination of factors which have already been outlined. The lack of effective continued customer support poses a major threat to the sustainability of the service, which had originally intended to support the most vulnerable businesses over time to remove the risk factors making them vulnerable, and is therefore likely to have an impact on the effectiveness of the scheme.

The management of BCD point out, however, that the appointment of dedicated area-based business crime prevention officers (which they have facilitated in Liverpool City Centre and hope to elsewhere) will enable continued customer care to be provided. Nevertheless, there have been no attempts to revisit or support individual beneficiaries to date.

The completion of an evaluation questionnaire will provide an opportunity to go some way towards fulfilment of this objective, since it will be the first time that recipients have been contacted by the service. It will therefore serve three functions. First, it will act as an independent evaluation of the experience of using the service. Second, it will enable any change in the respondent's exposure to and experience of crime to be explored, thereby offering an indication of the relative success of the security measures and advice facilitated by the service. Finally, it will provide an opportunity for the promotion of the service's new strategy and approach, for newly-produced advisory material to be made available, and for the generating of new leads for the potential development of group schemes. Hence the follow-up call will form a link between the old strategy (and its recipients) and the new cluster approach. Thus the questionnaire will not simply be a tool for evaluation, but will also explore the possibility of continuing the relationship between BCD and its users which should already be in place.

Additional activities

In addition to its core operations, the service has participated in a number of other schemes. BCD has held fraud and crime prevention seminars with the Police Fraud Wing, and was involved with Operation Guardian, which sought to reduce robberies against small businesses and cash-in-transit vehicles. It has assisted in the management of Lancsafe, which explored the occurrence of crime along the A580 'hot ribbon', and has successfully persuaded the police helicopter to overfly business crime hotspots when returning from its other duties.

BCD has also been successful in generating funding. It has acted as an agent for four Merseyside local authorities in the administration and delivery of funds accessed under the Funding for Retailers in Deprived Areas (FRIDA) scheme. It has also successfully accessed Partnership Development Fund (PDF) funding from the Government Office for the North West, and has delivered regeneration funding made available through the Single Regeneration Budget (SRB), Strategic Investment Area (SIA) and New Deal schemes across Merseyside. The service has also conducted consultancy work for Manchester Crime and Disorder Reduction Partnership to develop action plans for reducing business crime at selected hotspots across the city.

The way forward and recommendations

The actual strategy for the forward development of the BCD service remains unclear. Bids have been put forward to access Objective One funding, and the management of the service wish it to be used as a model which could be rolled out regionally or even nationally. The changed strategy has gained widespread support, despite the fact that its effectiveness is not yet known, as can be seen by the number of group schemes that have been implemented. The role of BCD has therefore become one of facilitation, support, and the provision of advice and technical support to clusters of businesses. This approach, in which BCD could act as an 'umbrella group' supporting networks of business associations, enables businesses to work together, as communities, to prevent crime.

Information-sharing

There remain major problems with the availability of the information which enables business crime to be effectively prevented. These have been outlined in a previous section of this chapter, and will not be repeated here. It is necessary simply to say that there is a shortage of information available from reported and recorded crime statistics, and that what information is available is of an *ad hoc* kind, and to a large extent lacks comparability.

Hence the role of BCD would be twofold: first, one of facilitator, nurturer and supporter of business clusters implementing group crime prevention schemes; and second, one of an information hub, ensuring that the best data and information are made available to ensure that trends in business crime are identified, and that advice is offered to ensure these are efficiently dealt with. Part of this latter role would be to collect data in a standardised format from each of those business clusters supported, to facilitate research into the environmental and procedural factors influencing business crime.

Lessons learned and recommendations

Although the evaluation of BCD is not yet complete, and its effectiveness in business crime prevention cannot yet be claimed, it has made substantial progress, not least in raising the profile of business crime and illustrating the eagerness of a range of partners to prevent such crime. Many lessons have been learned from the implementation of this project, and on the basis of their experience of involvement in BCD, the authors have a number of observations to make. In particular, they recommend that the following should be borne in mind when implementing a scheme of this type:

- *Marketing.* The advantages of employing marketing professionals to undertake the marketing of schemes should not be underestimated. This is particularly important, as research has found that publicity for crime prevention schemes alone has a significant effect on levels of victimisation, and can actually reduce crime independently of the implementation of physical measures.[17] It is also important to continue marketing the service so it is not forgotten by past users.

- *Customer care.* The scheme must ensure that it has the ability, in terms of the existence of a customer service infrastructure supported by the necessary information and staff knowledge, to deliver continued customer care. Happy customers will tell many other potential users of the experience they have had, thereby enhancing the impact of ongoing marketing campaigns.

- *Call centre.* When a call centre is employed, it is crucial to negotiate payment on a 'per call' basis, perhaps with minimum and maximum targets and set review periods. While the difficulties of estimating demand are acknowledged, different scenarios—which may include the redeployment of funding—should be considered.

- *Information overload.* The dangers of over-ambitious data collection should be remembered—only the data that is necessary to make the service operate most efficiently should be collected, though the IT system should allow for the collection of additional data items at a later date.

- *Information skills.* At least one of the executive officers should have skills and experience in dealing with information. Ideally, an information specialist should be one of the executive members of the scheme.

- *IT skills.* The budget should allow for the deployment of an IT/database developer who can work with the information specialist to develop and cost the necessary IT and information strategy, and who can act as a 'one-stop shop' for all the hardware and software requirements. This specialist should be involved from the pre-application stage.

- *Commitment to data provision.* In exchange for the crime prevention services offered by the service, the recipients must understand (perhaps by contract or in some similar fashion) that they are required to offer a certain amount of data about their business. This may be used to keep Chamber of Commerce or even NLPG databases up to date, but will also be made available to further research into the nature and prevention of business crime. The importance of data collection should also be communicated to all stakeholders of the service, perhaps by example.

- *Avoiding exclusivity in group schemes.* Group schemes offering advice and support, and the building of ongoing relationships with businesses, appear to be the most effective and efficient approach to business crime reduction.[18] These schemes may be offered as additional benefits membership of business associations or chambers of commerce, but the danger of excluding the most vulnerable should be borne in mind. By adopting a payment/pricing structure which reflects the vulnerability and resources available to each business for crime prevention, exclusion of those most at need may be avoided. Naturally, a member's status would need to be reviewed, to ensure fairness, each time fees are required (the Gold/Silver/Bronze approach).

- *Conceptualisation and implementation.* It is clear from the account above that for a number of reasons the BCD initiative did not achieve all its original objectives. Although the exact reasons for this remain unclear, the authors believe that the involvement of the implementation team in the bidding process (rather than recruiting the team's members once funding has already been obtained) would help to ensure that the scheme stayed as close as possible to its original objectives. This would allow members' ideas to be taken into account at this earlier stage, and increase their ownership and understanding of the initiative's original conception. Furthermore, if it is seen as important that an initiative is not allowed to stray from its original objectives, those involved in the conceptualisation of the scheme should continue to be accountable in some way for its running and development.

Conclusion

The operational development of BCD illustrates the difficulties that can be experienced when trying to devise a scheme which aims to derive maximum benefit in security provision from limited resources. The BCD scheme set out to manage and financially aid the delivery of security advice and equipment to a vast number of businesses across Merseyside. It was originally intended that the best way to select those businesses that were to become recipients was on the basis of their vulnerability. However, the service has naturally evolved an approach in which the limited resources are shared equally among as many businesses as possible, by supporting the development of group schemes. In this way the service is perhaps taking a more sustainable approach—as in community and social regeneration—where businesses are encouraged to work together and to share their own resources and efforts, and are supported in doing so.

When considering the achievements of the scheme, it is therefore important to bear in mind that a balance should be struck between any scheme's original objectives and its natural evolution. Current research has stressed the importance of seeing crime prevention in a similar way to the arms race: as offenders adapt and become aware of new risks, schemes should be modified to combat this effect.[19] The BCD scheme has shown signs of such evolution.

Notes

1 The authors all work in the Environmental Criminology Research Unit (ECRU), Department of Civic Design, University of Liverpool; email: douglass@liverpool.ac.uk.

2 Fisher, B. (1991) A Neighbourhood Business Area Is Hurting: Crime, Fear of Crime and Disorder Take Their Toll. *Crime and Delinquency*. Vol. 37, No. 3, pp 363–73.

3 At *http://www.crimereduction.gov.uk/toolkits/br00.htm.*

4 Mirrless-Black, C. and Ross, A. (1995) *Crime Against Retail and Manufacturing Premises: Findings from the 1994 Commercial Victimisation Survey*. Home Office Research Study No. 146. London: Home Office.

5 Bowers, K. (1999) *Crimes Against Non-Residential Properties: Patterns of Victimisation, Impact upon Urban Areas and Crime Prevention Strategies*. Unpublished PhD thesis, University of Liverpool.

6 The full list of partners includes: Merseyside Police; Merseyside Police Authority; the Safer Merseyside Partnership; Liverpool Chamber of Commerce and Industry, and other chambers of commerce on Merseyside; Southport Telematics; Central Southport SRB Partnership; and the Environmental Criminology Research Unit, University of Liverpool.

7 This term is used in marketing to define the maximum target population of a campaign.

8 At *www.businesscrimedirect.org.uk.*

9 The 'Integrated Criminal Justice System' (ICJS).

10 British Chambers of Commerce (2002) *Securing Enterprise. A Framework for Tackling Business Crime*. London, BCC.

11 Bowers, K.J., Hirschfield, A.F.G. and Johnson, S.D. (1998) Victimisation Revisited: A Case Study of Non-Residential Repeat Burglary on Merseyside. *British Journal of Criminology*. Vol. 38, No. 3, pp 429–52.

12 These are the areas designated as 'Pathway Areas', towards which European Objective One ERDF and ESF funding is targeted.

13 See for example British Chambers of Commerce, op cit.

14 Mostly those identified locally by previous research; see Bowers, op cit.

15 Townsley, M. and Pease, K. (2001) Hot Spots and Cold Comfort. In Tilley, N. (ed.) *Evaluation for Crime Prevention*. Vol. 13, *Crime Prevention Studies*. New York: Criminal Justice Press.

16 An example of this for residential burglary can be found in Bowers, K.J., Johnson, S.D. and Hirschfield, A.F.G. (in press) *Pushing Back the Boundaries: New Techniques for Assessing the Impact of Burglary Schemes*. Home Office Research Series No. 246. London: Home Office.

17 See for example Johnson, S.D. and Bowers, K.J. (submitted) Opportunity is in the Eye of the Beholder: The Role of Publicity in Crime Prevention. *Criminology and Public Policy*. Vol. 2, No.3. See also Laycock, G. (1991) Operation Identification, or the Power of Publicity? *Security Journal*. Vol. 2, No. 2, pp 67–71.

18 This claim cannot currently be substantiated, although the full evaluation of the scheme will attempt to do so.

19 Ekblom, P. (1997) *Gearing Up Against Crime: Can We Make Crime Prevention Adaptive by Learning from Other Evolutionary Struggles?* Paper presented at the 6th International Seminar on Environmental Criminology and Crime Prevention, Oslo.

Chapter 3

Loss Prevention: Senior Management Views on Current Trends and Issues

Read Hayes[1]

Large retailers employ senior executives to manage multi-billion-dollar crime and loss issues. This chapter examines some of the issues and trends confronted by these important managers, in order to inform current and future loss prevention managers. Some of the topics studied include the changing mission of loss prevention executives, loss prevention performance measures, current and future crime and loss issues, how loss prevention executives evaluate loss control technologies, and their sources of loss prevention best methods. The study consists of information provided by a sample (n = 60) of senior loss prevention managers from 59 US retail organizations. Loss prevention managers reported they prefer using shrinkage measures to numbers of apprehensions to gauge performance. They also frequently go into partnership with store operations managers, but rarely with merchandise buyers, on loss control issues. Finally, the participants indicated that dishonest staff was the most pervasive issue they currently dealt with, and that they did not expect this to change in the future.

Introduction

Retailing is a large and vital part of the world's economy. After services, it is the largest employer in the United States, accounting for over 21 million American jobs, or 18 per cent of non-agricultural employment. Mega-retailer Wal-Mart, with more than $190 billion in annual sales and over 3700 stores, is now the largest United States employer—it has over 800,000 associates, and the number is growing. Retail sales in the United States easily exceed $1.54 trillion annually.[2] Further, the dynamics of retailing continue to change. Bamfield and Hollinger estimate that in 1993–1994 error and thefts by staff, customers, and others throughout the supply chain cost retailers in both the United States and the United Kingdom between 1.6 per cent and 2.0 per cent of their sales—or over $30 billion per year.[3]

Retail crime and loss control is a complex, and vital, part of retail operations. Inventory losses tend to cluster around specific locations and highly desirable items, reducing consumer availability and sales. Another major concern for retailers is that fear of crime will result in customer avoidance behaviour, such as limiting shopping activity to daylight hours, shortening shopping visits, and switching to competing retailers, or to shopping formats such as the web, catalogues or television.[4] Retailers must also work to protect losses of proprietary information, cash receipts, physical assets, and even their reputations through catastrophes such as product tampering, civil unrest, and natural disasters. Another growing problem facing US retailers is constant legal claims. Customers and staff often sue if they feel they were wrongly accused of theft, if their physical apprehension or sanction was mishandled, if they were involved in on-site or work-related accidents, or if they were injured or killed during a crime incident.[5]

Due to the scope and complexity of these issues, retail companies very often employ senior executives to exclusively focus on crime and loss control.[6] These managers provide asset protection problem analysis, strategic planning, budgeting, and operational expertise to their companies.[7] Of current interest is how these managers are oriented, what they focus on, how they gauge their own performance, and what they consider when making operational decisions. This exploratory study is designed to provide some systematic evidence to help answer these and other questions about loss prevention executives.

Background and research focus

A review of the literature reveals most of the information on this topic is within non-academic trade magazines (eg *Security, Security Management, International Security Review*) and numerous how-to books.[8] This lack of rigorously obtained data on loss prevention management remains a problem—and an opportunity.[9] Of particular interest in the current study are some of the dynamics of loss prevention management. Crime and loss issues are a problem for most retailers, and many firms employ special loss prevention executives to handle them. But do these managers focus exclusively on security/loss prevention, or have they taken on new expanded duties ('mission creep')? Murphy and Criste provide limited insight into the diversity of issues corporate security managers deal with, including white-collar (employee) crime, workplace violence, information leakage and theft, and insurrection.[10] While the authors were primarily surveying international security managers (n = 18), interestingly they mention general theft as a very small concern of the participants. No participant demographics are provided, so possibly retail operations were not included in the sample. Manunta, in a paper on the concept of security and loss prevention, also mentions that loss prevention goes beyond simple crime prevention,[11] as do Hollinger *et al.*[12]

Another area of interest involves loss prevention performance measures. In earlier times security managers seemed oriented primarily toward achieving high 'body counts', where the more shoplifters and dishonest employees apprehended, and the more stolen goods and cash recovered, the more effective the program. One possibility for this orientation was naivety on the part of very senior company executives such as CEOs and CFOs. These senior executives might have recognized the obvious problems of thieves, and expected that the more removed on a monthly basis, for instance by plainclothes store detectives, the fewer incidents of theft and violence they would have to hear about. But this focus may have changed, with loss prevention people, programs and systems designed to affect crime event and loss levels more proactively by actually preventing crime and loss events.[13]

No man is an island, and no successful loss prevention person can affect his/her company's performance single-handedly. This chapter is also interested in who loss prevention managers tend to go into partnership with in preventing crime and loss problems. Some security managers seek out and engage with outside law enforcement, consultants, and government officials,[14] but loss prevention also works with groups within its own organization to address company crime and loss. Another question of interest is what the current and future problems are which loss prevention managers are dealing with, and expect to in the near future. Perceived problems drive loss prevention actions.[15] Dishonest or error-prone employees, customers, or suppliers, and third-party visitors (hackers, robbers and burglars), combine to provide varied challenges.[16]

Loss prevention managers are busy people dealing with the myriad issues mentioned above, so prioritizing their time is an important topic. Organizational change, personnel and budgetary issues, executive protection, crime investigations, and dealing with natural and man-made disasters are examples of work tasks which loss prevention managers routinely perform.[17] This chapter also examines the time allotted by loss prevention executives to various workplace activities. Just as they must partner others in tackling the complexities of crime and loss, they also use technology to detect, prevent, and document constantly evolving problems.[18] Of interest in this project is what criteria loss prevention managers use when deciding to acquire two specific technologies: closed-circuit television (CCTV), and electronic article surveillance (EAS). Both CCTV and EAS are designed to deter theft activity by increasing the probability of detection and sanction for would-be thieves.[19] Finally, since decisions about prioritising, preventing, and handling crime and loss are critical, this study examines the sources of information which loss prevention executives use to inform their decisions. Often, loss prevention decision-makers are forced to rely solely on anecdotal or weakly derived research data before deciding what actions to take. There have been increased anecdotal reports to the author about senior retail company executives demanding decision-making information superior to pure anecdote, and to simple benchmarking with other 'successful' retailers.[20] In order to provide information on these elements of security management to loss prevention executives and scholars, this study examines all of these issues with a sample of senior retail loss prevention managers.

Method

Procedure and participants

The subjects in this study consist of busy executives who tend to travel extensively, making research contact difficult. Project data came from a telephone survey conducted over a 32-day period. A survey call team of students was trained and supervised by a PhD candidate to schedule 10–15 minute phone survey sessions with all loss prevention executives on a prepared call list.

The study sample was composed of 60 senior retail loss prevention executives from 59 companies. The participants were contacted from a list of 119 such executives who attended the National Retail Federation's or International Mass Retail Association's 1999 or 2000 loss prevention conferences. Of the initial list of 119, 41 did not respond, and 18 refused, for a response rate of 50.4 per cent (n = 60). As shown in Table 1, the majority of the respondents were very senior managers with titles of Vice-President (n = 20) or Director (n = 31), with only nine subjects (15 per cent) describing themselves as managers or other (such as a senior supervisor).

Table 2 indicates that most of the participants (n = 36) worked for specialty retailers such as books and recorded media, home furnishings, sporting goods, furniture, children's apparel, and jewellery. Eleven department stores participated (18 per cent), along with two mass merchants (three per cent), and 11 companies (including catalogue showrooms, close-out merchants, electronics, and paint stores) which described themselves as 'other store type' (18 per cent).

Instruments and limitations

The questions developed for this project were derived from discussions held during three annual loss prevention brainstorming sessions held in Orlando, Florida, from 1995 to 1998.[21] Over 100 senior loss prevention decision-makers, from more than 80 companies and based in several countries, participated in these problem-solving conferences. A 13-item questionnaire, designed to be short enough to allow very busy executives to participate, was developed from the original questions taken from the results reports of the conference sessions to the loss prevention participants. This instrument was designed to capture demographic information on the participants, along with data about their job responsibilities, performance measures, internal and external partnering dynamics, and decision-making resources and cues. All responses were generated from a review of the BrainStorm Conference reports, and trade and peer-reviewed loss prevention and security literature. The initial question on the executive's responsibility was a 'yes-no' response; the question on the amount of estimated time committed monthly to a certain area of responsibility had five ranges of hours ('less than 10 hours per month' = 1, to 'more than 25' = 5); while the balance of the questions used four-point Likert-type scales (ranging from 'very important' = 4, to 'very unimportant' = 1) to gauge perceived commitment or the significance of the issue.

Table 1. Job titles of participating executives

Job title	Frequency	Percentage
Director	31	52
Vice-President (Company Officer)	20	33
Manager	8	13
Other	1	2

Table 2. Type of retail company

Retailer type	Frequency	Percentage
Specialty	36	60
Department store	11	18
Other store type	11	18
Mass merchant	2	3

As with any research, this project suffers limitations imposed by its design and execution. The sample of loss prevention executives, given its size, is representative of the population and of the diversity of participants by type of retailer. However, generalizing the current findings beyond the sample has risks, since it was not a large or randomly selected group. The questions and response were derived from a large group of retailers, and from the current loss prevention literature, but may limit the range of responses. Likert-type scales provide a fair, but not a precise measure of intensity. The students used to collect the data were trained and were asked to be consistent in their survey-taking, but there may have been some differences in ability and execution. The analysis of the data, while informed by the literature, prior research, and experience, is considered subjective.

Results

The initial questions on company functions or operations for which the participating loss prevention executives have responsibility were combined into a single table and listed in rank order according to the frequency of a 'yes' response. Responses to the Likert-type survey questions were listed in rank order according to the mean score of the response. The standard deviation (SD) of each response is also listed, to indicate the groups' level of agreement.

Loss prevention executive functions, performance measures and partnerships

Table 3 indicates that all participants are in charge of loss prevention; 72 per cent of them also oversee company safety. Internal audit falls under some participants' area of responsibility, while 38 per cent manage inventory control. Some 25 per cent of the participants work on insurance claims management, but only eight per cent actually procure outside insurance coverage (or set up self-insurance). Other areas covered by the study's loss prevention executives includes: two who oversee their company's human resource function, two who manage inventory distribution, and one who oversees store operations as well.

Table 3. Functions for which loss prevention managers are responsible

Job area function	Percentage	Frequency
Loss prevention	100	60
Safety	72	43
Internal audit	45	27
Inventory control	38	23
Insurance claims management	25	15
Insurance procurement	8	5
Other	32	19

Listed under 'other' functions, and with single respondents per company activity, were: price changes, e-commerce, transportation, and employee screening. Many of the sample indicated they were also responsible for the physical security of all corporate assets and facilities. Almost two-thirds of the subjects now supervise safety and other non-security related functions, and almost two-thirds are in charge of coordinating customer and employee safety programs. Likewise almost half of the sample manage internal audit functions. Auditing often falls under the finance department; inventory control is found either within finance or in the commercial or merchandise buying departments. Both of these functions attempt to control loss by implementing and maintaining sound processes and control procedures—so the loss prevention fit makes intuitive sense. Many loss prevention departments realize that loss is a function of operational, demographic, process, and execution issues, not just purely a crime problem.

Like safety, insurance purchasing and claims processing are related to loss prevention. Managing risks and probabilities is often a part of preventing losses; and since losses include cash, legal and insurance claims, and premium increases due to high risk or incident levels, loss prevention is uniquely positioned to synthesize these operations. Less clear are the inclusion of transportation, e-commerce, and price changes. These operations are logistical or marketing in nature, but seem to be particularly high-risk processes, and loss prevention departments with networks of loss prevention specialists and investigators can design-in protective functions, and provide protective manpower. Personnel screening for the risk of workplace deviance also seems a good fit with loss prevention. But loss prevention managers must remain focused on the overall company goals of hiring productive, honest people who work well with others.

Retailers are in the business of selling merchandise, and the unexplained loss of merchandise, or 'shrinkage', was rated as the most important measure of loss prevention departmental performance (Table 4). This measure also had the lowest standard deviation (0.47) indicating the highest level of participant agreement. While apprehended offenders was rated last, its score (3.13, SD 0.70) was very close to shrinkage rates as a performance measure (3.85), indicating its continued existence in the face of anecdotal rejection as a meaningful indicator of performance success by many retail operations.

Table 4. Importance of loss prevention departmental performance measures

Performance measure	Mean Likert score	SD
Recorded inventory loss or shrinkage	3.82	0.47
Feedback from store managers	3.78	0.56
Reported crime and loss incidents	3.35	0.88
Apprehended offenders	3.13	0.70

The manager feedback metric had a mean rating of 3.78, and an SD of 0.56, since many loss prevention executives consider the store and distribution centre managers to be their 'customers', striving to provide them with high-quality crime and loss control support. The third-ranked performance measure, at 3.35, was the level of reported crime and loss incidents. This metric had the highest standard deviation, 0.88.

The participants rated shrinkage or item loss measures as the most important indicators of loss control success. Violence and other crime issues occur, but retailers exist to market other people's products, and if those products are stolen or lost, lower sales and profits result.[22] The idea of shrinkage appears simple: the difference between what the records show should be in the store (based on item shipments, less recorded sales) versus what a physical count indicates is actually on-hand; the difference between expected and actual (usually a loss rather than a gain) is often stated as a percentage of sales.[23] In the field, however, inventory is very difficult to manage and count. Large numbers of stores and distribution centres spread out over vast nations, and the world, create a massive and complex system based largely on assumptions and containing inherent error. Retailers increasingly rely on only partial physical counts (which have their own error issues), and more on item adjustments made at the store level, or by buyers at a central office. As these measures become more complex, they might lose credibility with field managers who are able to see discrepancies between what the computer says is on-hand and what they are seeing.

As expected, Table 5 indicates survey participants rated store operations as their primary business partner (3.85, SD 0.48). Store operators are the primary 'customers' for loss prevention efforts such as security procedures, store detectives or specialists, investigators, loss awareness campaigns, and control technologies such as electronic surveillance tagging, point of sale, alarms and cameras. Inventory control was the second highest-rated partner, at 3.75 (SD 0.51) due to its work to accurately move and track merchandise throughout widespread systems, while human resources were the third-rated partner, at 3.68 (SD 0.54). Human resources work with loss prevention on pre-employment screening, investigations, and employee morale and performance. Internal audit, which works to ensure stores and other departments are using transactions and procedures properly, had a mean score of 3.53, and the highest SD, 0.89. Despite its criticality in co-developing decision-making data programs for loss prevention, computer support or IT had only the fifth-highest rating, of 3.35, and an SD of 0.68.

Distribution centres (DCs) had a mean rating of 3.35 and an SD of 0.68; finance had a mean of 3.33 and an SD of 0.63. Surprisingly, outside loss prevention suppliers or vendors were rated above company buyers, with a mean of 2.82 and the second-highest SD, 0.87. Notwithstanding their role in buying, and pushing appropriate levels of sellable merchandise to the right stores, buyers were last, with a rating of 2.78 and an SD of 0.80.

Table 5. Preferred partners for the loss prevention department

Action partner	Mean Likert score	SD
Store operations	3.85	0.48
Inventory control	3.75	0.51
Human resources	3.65	0.54
Internal audit	3.53	0.89
Information technology	3.52	0.60
Distribution centres	3.35	0.68
Finance	3.33	0.63
Outside loss prevention suppliers	2.82	0.87
Merchandise buyers	2.78	0.80

Loss prevention issues

Table 6 illustrates that there is no difference between the priority ranking of loss prevention issues the participants face now and those they expect to face over the next five years. Not only are the mean Likert score rankings similar, but so are the SDs.

Table 6. Relative importance of current and future loss prevention issues

Loss prevention issue	Current mean score	SD	Future mean score	SD
Employee theft	3.95	0.22	3.97	0.18
Customer theft	3.48	0.57	3.47	0.62
Online computer crime	2.82	0.79	3.33	0.71
Parking lot and other violence	2.72	0.85	2.93	0.78

Employee deviance (especially theft of assets) was ranked as the loss prevention issue having the highest importance (current 3.95, future 3.97). Employee deviance consists of all those workplace behaviors that are prohibited and harmful to the organization. They range from tardiness, not following procedures, theft, and harassment, to violence. This issue also had the most agreement as to its importance, as indicated by its very low SD (current 0.22, future 0.18). This finding indicates that pre-employment screening and procedural compliance enhancements are paramount for security managers. Likewise, customer theft is now (3.48, SD 0.57) and will be in the future (3.47, SD 0.62) considered the next most important issue. Shoplifters tend to strip stores of their most desirable items, costing sales, and evidence-based deterrence measures will continue to be important. On-line computer fraud (current 2.82, SD 0.79, future 3.35, SD 0.71), such as credit card application fraud or website attacks, was considered slightly more important than parking lot and other violence in retail settings, although the future mean scores given for these two issues increased slightly over those for the current ranking, and their SDs fell as well. Senior loss prevention executives need to stay up to date on computer security issues. Even though violence control was the lowest-ranked issue area, employees' and customers' perceptions of their own safety are critical to maintaining revenues.

Contrary to the title of loss prevention, Table 7 indicates that responding to, rather than preventing, issues such as large employee theft incidents or violence in particular stores is the top-ranked item (3.43) and had the second-lowest SD (1.27). Tactical operations rather than strategic analysis and planning tend to dominate senior loss prevention managers' time. Likewise, immediate personnel matters were a close second with a rating of 2.98 and an SD of 1.65. Investigation management had the third-highest mean score, 2.90, with an SD of 1.54.

Table 7. Amount of time spent monthly on select work issues

Operational issues	Mean Likert score	SD
Responding to current issues	3.43	1.27
Loss prevention personnel/staff matters	2.98	1.65
Investigations management	2.90	1.54
Building working partnerships	2.80	1.31
Long-term loss prevention planning	2.47	1.33
Loss prevention budget planning and management	1.77	1.13

Building partnerships with fellow managers and outside suppliers and experts came next, with a mean of 2.80 (SD 1.33). Strategic planning came in second-last, at 2.47 (SD 1.33), followed by budgeting and financial management, at 1.77 (SD 1.13). Immediate issues are important and time-consuming, but it may also be easier for many managers to spend most of their time reacting to problems, talking to subordinates and attending meetings rather than investing more of their efforts in the systematic analysis, planning and executing of strategic measures to position their department and organization for high performance over time.

In addition to the responses shown in Table 7, 23 participants provided additional, critical loss prevention issues they felt should be addressed, including: injuries and claims on company property; vendor/delivery theft and fraud; operational, distribution, system, and administrative errors which create losses; liability for actions or inaction on company property; credit card and check fraud; cash control and other procedural compliance; civil disorder on company property; and maintaining good customer service.

Loss prevention decision-making
Chain retailers make capital investments in differing types of technologies designed to increase revenue and profit. Two common and important surveillance technologies were chosen for examining and prioritising the criteria which loss prevention decision-makers use to select a particular type or brand of technology (Table 8).

The ranking of buying criteria were similar in both technologies, with the exception of product appearance, brand name, and training support. Reliability of the product was ranked at the top for both systems. Also comparable in ranking were: the system's performance, service availability, ongoing or operating costs, initial product purchase or lease price, ease of system use, and salesperson relationship.

Table 8. The EAS and CCTV buying criteria

	EAS		CCTV	
Criteria	**Mean Likert score**	**SD**	**Mean Likert score**	**SD**
Product reliability	4.07	0.45	4.08	0.59
Product performance	4.00	0.52	3.93	0.71
Service availability	3.93	0.58	3.90	0.75
Ongoing product costs	3.92	0.67	3.85	0.80
Product price	3.92	0.59	3.82	0.81
Ease of use	3.82	0.65	3.70	0.99
Salesperson relationship	3.47	0.89	3.35	1.08
Product appearance	3.40	0.92	3.33	1.05
Supplier name brand	3.32	0.97	3.20	1.05
Training support	3.30	1.08	3.15	1.09

The physical appearance and aesthetics of EAS systems were ranked as a more important issue than they were for CCTV (EAS 3.40, SDS 0.92; CCTV 3.15, SD 1.09). Not measured here is whether loss prevention executives believe CCTV domes and EAS tags, deactivators and pedestals should be more visible to boost deterrence, or whether the systems should simply look better. The supplier's brand name was the second-lowest ranking for both loss prevention technologies (EAS 3.32, SD 0.97; CCTV 3.20, SD 1.05), indicating suppliers must strive to market the actual, comparative efficacy of their systems over those of competitors. An interesting dynamic with these large systems is that their relatively high capital costs and widespread installation across national chains may strongly influence perceived brand loyalty. The diversity of opinion on buying criteria appears to broaden as the mean score lowers, according to the SDs.

An important component of security management at any level is decision-making; and aside from the decision-maker's personal abilities, the quantity and quality of decision-making data is the most important part of this activity. This is an area of loss prevention management that deserves priority treatment. While the rankings were fairly close on the importance of decision information sources, the executives in the current sample rely most heavily on their peers, both inside and outside their companies, for advice and referrals (3.67, SD 0.60). The quality and source of information provided by peers was not examined here, but is an important topic, particularly with the possibility of circularity generated by decision-makers passing on the same ideas and beliefs in a closed loop of conferences and magazines that do not require peer-reviewed critiques of the information presented and how it was derived. The study's participants also rely heavily on internal data generated by their staffs and company systems, such as point of sale (POS or Epos) and shrinkage estimates (3.43, SD 0.67). Studies conducted outside their companies (3.12, SD 0.52), and loss prevention conference workshops put on by retail trade organizations (3.12, SD 0.61), were tied for the third-highest ranking.

Table 9. Primary sources of loss prevention decision-making information

Information source	Mean Likert score	SD
Other loss prevention professionals	3.67	0.60
In-house loss prevention research/data	3.43	0.67
External loss prevention research	3.12	0.52
Loss prevention conference workshops	3.12	0.61
Other loss prevention training workshops	2.93	0.52
Loss prevention product suppliers	2.92	0.56
Trade magazines and journals	2.92	0.62
Loss prevention textbooks	2.58	0.70

Outside workshops and seminars on specific topics, such as offender interviewing techniques and loss prevention management, had a mean score of 2.93 (SD 0.52). Loss prevention product suppliers (2.92, SD 0.62) also provide information on both their products and those of their competitors. They also disseminate industry and general loss prevention information through sponsorship of loss prevention studies and reports. Finally, trade magazines (2.92, SD 0.62) and textbooks also provide loss prevention information (2.58, SD 0.70).

As with earlier questions, participants were encouraged to share other relevant ideas. Five participants listed other important information sources, including law enforcement, online databases, and the Internet in general.

Discussion

Loss prevention texts mention the responsibilities of loss prevention managers in broad terms, but tend to focus solely on the prevention of theft and violence. But discussions at the BrainStorm loss prevention conference series indicated a slow evolution towards including other areas in the loss prevention portfolio (loss prevention mission creep). This chapter supports that contention. What is not clear is over what period this change has been or is occurring, and why. There are at least two possible explanations: organizations continue to reduce the size of their executive staff, which results in fewer executives doing more; or possibly certain loss prevention executives are particularly effective, proven managers ready and able to take on more responsibility. There is a danger, however, that the focus of current loss prevention executives will become too broad, resulting in poor execution in one or more areas—particularly those in which the manager is not as interested or skilled. Also, successor managers may not have the same skill sets to manage a more diverse department.

As anticipated, using the number of apprehended thieves as the primary indicator of loss prevention success was found to be the least favourite choice. Traditionally, retail security and loss prevention managers tried to maximize the number of apprehensions of dishonest employees and customers, both because these people create the company's loss problems, and because their numbers are easily tracked. While apprehensions provide some deterrence, and are necessary (particularly in high-theft stores), the 'body count' performance criterion seems more reactive than proactive.[24] Also, apprehensions themselves come at a high price in the cost of security staff people and technology, as well as creating legal risk exposure.[25] And this author has found no evidence in other data that apprehension levels predict loss levels.

Other criteria such as recorded loss levels and store manager feedback may provide more meaningful information as to mission accomplishment. Loss prevention managers usually value the perceptions of good store and regional managers regarding the execution and effectiveness of loss prevention initiatives. This measure is subjective at best, but well-briefed managers can provide valuable feedback for action. Accurate company and store loss levels help determine the effectiveness of overall loss control, but field experiments and rigorous correlational studies would provide better program impact data. It is also important that loss prevention managers can only control so much, meaning loss levels can be affected as much if not more by store operators. Reported crime and loss incidents were also indicated as valuable, but many companies do not consistently or accurately record all or even most incidents in the stores or parking areas.

Since loss control is a multi-faceted and complex problem, responsibility for its operation cuts across departmental lines. Protecting people and physical assets is the responsibility of all employees, and loss prevention executives rely on and work with intra-company partners on loss reduction. Store managers, and store operations executives, are loss prevention departmental customers, and partners in the fight to suppress crime and loss levels. Inventory control managers provide valuable information on merchandise processes, flow, and risk areas. Human resource executives work with loss prevention to screen out potential thieves or overly aggressive staff, and also jointly work on creating a positive, ethical workplace that can reduce theft and assist in internal investigations.[26] Internal audit helps loss prevention maintain procedural audit compliance in ever-increasing, chain-wide operations. This is vital, since work process failures are a primary cause of loss, and create vulnerabilities for theft and fraud. Information technology (IT or IS) builds software to track product and cash transactions and movements in order to provide reporting on emerging theft or incident patterns, as well as indicators of dishonesty throughout the supply chain. These systems can track asset flow from the suppliers, through the distribution centres, to the point of sale and return to vendor, with exception reporting and pattern recognition applications. Finance staff provide analysis expertise on the financial return on planned and existing loss prevention investments. They can also help loss prevention interpret loss data in financial impact terms in order to prioritise operations.

Outside suppliers of loss prevention technology and consultants provide unique and outside perspectives and solutions for loss prevention executives. Somewhat of a surprise was that merchandise buyers (merchants) were the least-used partners. Buyers should be tracking the sales and loss performance of their items throughout the supply chain. They have a vested interest in putting and keeping all of their items on the shelf and available for purchase. They should also be looking for the best, most cost-effective ways to protect item availability through procedures and technologies. The participants also deal with law enforcement authorities and loss prevention staff from other retail companies on a regular basis, to track high-rate offenders and identify hot spots for concentrated action. Security managers should include plans to enter into partnership with those most likely to help them control crime and loss in their program and strategy.

As discussed, loss prevention managers deal with a large number of issues. The top issues the participants deal with now, and will in the future, turned out to fall into the same order in the current study. Employee theft was the term used, but is routinely extended to include workplace deviance. Employees daily spend long periods of time on the job, and bring with them, or create, many motivations for dishonesty and other rule-breaking in the workplace.[27] They may steal product, cash, information, and time.[28] Employees can also interrupt the business by aggressive behaviours toward other staff, and customers. This topic deserves more research, perhaps looking at how loss prevention decision-makers determine the relative importance of each area. Is there a formula that takes into account the relative frequency of past events, plus the criticality or severity of the damage to which particular types and causes of crime and loss contribute?

Shoplifting was ranked the number two issue, and according to the participants remains a problem. Shoplifters are a major cause of inventory loss, and these loss levels have remained relatively stable, in a range between 1.3 and 1.9 per cent of gross sales, over the last decade.[29]

Loss prevention executives are busy people, but are not necessarily spending most of their time on preventing problems—the participants indicated they actually spend more time responding to them. Large thefts, widespread item loss problems, inquiries from their superiors, violent incidents, and legal claims were mentioned as some of these types of issues. The participants also revealed that they tend to spend a large amount of time dealing with their own staff (ie recruiting, hiring, training, transferring, disciplining, and terminating employees): not normally a proactive action, but important for keeping their operations on track.

The third-ranked issue was handling investigations management. Employee deviance can be confined to a single person committing a solo theft, or can be found to be widespread and containing multiple types of problems, such as timecard fraud, cash theft, and merchandise theft at the same time. Evidence must be complete and credible, as also should reports—there are lessons to be learned from these incidents.

Not all of a loss prevention executive's time is spent responding or reacting; the study's participants indicated they invested their time in building relationships and partnerships, as well as in conducting some long-term analysis and planning. They also must prepare detailed financial budgets, and be prepared to explain and defend each and every detail of the proposed plan and budget. The area of time management and prioritization for loss prevention managers is ripe for further review. Indications of the percentage of time spent understanding problems and testing optimal responses, as compared to operating the initiatives, would be valuable to an understanding of loss prevention management.

Loss prevention executives also play a strategic role in their companies by specifying, testing, and recommending loss prevention systems. These products can range from small low-tech items such as signage, stickers, 'keepers' (which attach to merchandise to increase its size in order to decrease its concealability), and bars which keep shopping carts or trolleys inside stores, to large-scale deployment of EAS and CCTV systems in stores and distribution centres. Some of these purchases constitute large capital expenditures, and can require great selling skills to have them funded by senior company executives.

It was interesting that a system's reliability was slightly higher-ranked than its actual performance. These two items could have been confounded, or the retailer could have considered the efficacy of the systems high, and now simply wants them to operate as advertised over the long term. In the future, how retailers define and measure loss prevention program performance should be examined. Options include rigorous field experiments and regression analysis combined with financial performance metrics that would be preferred over anecdotal or poorly designed field tests. Service availability was very important, but training support was not. This seems surprising, since so many loss prevention systems and programs are never as simple or hands-off as they might first appear, and consistent system execution is critical to achieving annual objectives. Both initial and ongoing costs were important in selecting the right product, but with both EAS and CCTV ongoing cost was rated slightly ahead of initial purchase and set-up price. At this point it appears price is important, but ongoing effectiveness is considered more important in the selection process. Salespeople are important in the equation, but only moderately so considering the other factors. Like training support, the brand of the product did not appear as important as the product's performance.

The preceding section provided information on the criteria considered important for selecting a particular loss prevention technology. This section examines the sources of decision-making data used by loss prevention executives, for a wide range of issues such as determining and prioritizing company-specific issues, and selecting the best mix of loss prevention actions. The top item was other loss prevention professionals. Anecdotal reports of effectiveness are common in many professions. Medicine operated in a similar fashion for many decades, before switching to a systematic or evidence-based model.[30] The current study shows some movement in that direction, with in-house and external research findings rated in second and third positions respectively according to their mean ratings. The quality of corporate research probably includes poorly designed 'one-shot' projects, where one or many measures are tried in a single location, or with a specific product, and any observed effects are automatically presumed to be both causal in nature and generalizable to other locations and situations. Likewise, many evaluations suffer from other poor designs, such as selecting extreme cases (highest-loss stores, etc) and then noting the expected reduction (often the regression to the mean phenomenon), or confounding the effects of multiple actions taken simultaneously, and subsequently concluding that the loss prevention item of interest is the cause of any observed changes. Sound research should include proper sampling, measurement, control, and analysis to help loss prevention managers determine causality versus mere covariance, and to generalize their findings.

Other sources of decision information were loss prevention workshops at conferences (usually mounted by retail associations), training workshops conducted by outside trainers or consultants, loss prevention product suppliers who provide insight into both their own products and those of their competitors, and finally, loss prevention and security magazines and textbooks. Often these data sources do not dispense evidence-based data or recommendations, but rather base them on anecdote or conventional wisdom.

Future research

This project was designed to provide initial information on loss prevention management, and as such indicated some current patterns, and opportunities to learn more. The areas of responsibility of loss prevention executives are open for more investigation. Other questions would be how long this trend toward generalization from specialization for loss prevention senior managers has been underway, and for what reasons.

The topic of how loss is created in retail operations deserves much greater study. Loss stems from intentional and unintentional actions and omissions, and occurs throughout the supply chain. A combination of case studies and quasi-experimental projects can help loss prevention more precisely to identify the relative contribution to theft of customers and staff, to discover how much error is due to system/ process and how much to individual employees, and to locate problems at particular points within the supply chain.

All of the metrics which retailers use to evaluate loss prevention departments, individual loss prevention operatives, and loss prevention systems and programs are open to further examination. The adequacy and accuracy of these measures, their relevance to overall company goals, and how they are analyzed and used to develop loss prevention initiatives are areas of tremendous opportunity for further research. New performance metrics, such as loss by item per square foot, or by category, or by cost of goods, can be used for analysis. All measures should be relevant, and using rates such as these, and the traditional percentage of total or item sales, can help loss prevention. But other significant influences (such as gross margin) should be controlled for, to better isolate non-random clustering of loss by item, category, location, offender type (age, gender, etc), store procedural compliance and time, and interactions of these variables. This improved understanding would better inform prevention actions.

To be determined in other research might be how, and how rigorously, loss prevention managers evaluate the efficacy of loss prevention procedures, programs and systems. Items of interest include the metrics they use to determine how well these loss prevention initiatives affect loss or incidents, and their use of sound sampling, measurement, controls, and statistical analysis—all designed to determine the value of the loss prevention actions, and of the research itself. Further research in this area might include a look at the dynamics of the buying and in-house funding of expensive loss prevention systems. Prior to making buying recommendations, it is important to understand what criteria loss prevention managers use to evaluate the efficacy and financial value which specific loss prevention systems provide.

Conclusion

Loss prevention managers play an important role in their organizations. They directly affect sales and profit by managing key item availability, employee and customer safety, and cash and other asset protection. However, their role in their companies continues to evolve: many are being assigned a broader scope of responsibility, and dealing with non-crime issues that may ultimately dilute their loss control effectiveness. Loss prevention executives also appear, primarily, to be tactically rather than strategically focused, and frequently team up with others to address the myriad issues they are faced with—though they tend to work with operations and finance managers more frequently than with merchandise buyers. These managers also look to inventory loss measures (of varying quality) as a better indicator of performance than apprehension totals, or service surveys by store managers, but appear to look more often at overall measures rather than specific loss prevention efforts. It is clear that the quality of management decision-making may be adversely affected by a lack of quality, evidence-based data. Finally, the important job functions of loss prevention managers deserve more examination, since prevention activities are so critical to successful retail operations.

Notes

1 Read Hayes is a Research Associate at the University of Florida, and Director of the Loss Prevention Research Council; email: readhsd@earthlink.net.

2 National Retail Institute (1996) *Retail Industry Indicators*. Washington, DC: NRI.

3 Bamfield, J. and Hollinger, R.C. (1996) Managing Losses in the Retail Store: A Comparison of Loss Prevention Activity in the United States and Great Britain. *Security Journal*. Vol. 7, No. 1, pp 61–70.

4 Warr, M. (2000) *Fear of Crime in the United States: Avenues for Research and Analysis*. Vol. 4, *Measurement and Analysis of Crime and Justice*. Washington, DC: US Department of Justice, Office of Justice Programs, National Institute of Justice.

5 Federal, R. and Fogelman, J. (1986) *Avoiding Liability in Retail Security.* Atlanta, GA: Stafford.

6 Hayes, R. (1991) *Retail Security and Loss Prevention.* Stoneham, MA: Butterworth-Heinemann, pp 36–45; Jones, P.H. (1998) *Retail Loss Control.* 2nd edn. Oxford: Butterworth, pp 21–4.

7 Hayes, op cit.

8 Ibid; Jones, op cit; Neill, W.J. (1981) *Modern Retail Risk Management.* Sydney, NSW: Butterworth; Purpura, P.P. (1989) *Modern Security and Loss Prevention Management.* Stoneham , MA: Butterworth, pp 111–4; Pupura, P.P. (1993) *Retail Security and Shrinkage Protection.* Stoneham, MA: Butterworth-Heinemann.

9 Gill, M. (1996) Risk, Security and Crime Prevention: An International Forum for Developing Theory and Practice. *International Journal of Risk, Security and Crime Prevention.* Vol. 1, No. 1, pp 11–17.

10 Murphy, G.M. and Criste, B.F. (1997) A Study of Contemporary International Security Management: Results From a Small and Round Table Discussion Survey. *Security Journal.* Vol. 8, No. 3, pp 263–75.

11 Manunta, G. (1999) What Is Security? *Security Journal.* Vol. 12, No. 3, pp 57–66.

12 Hollinger, R., Dabney, D. and Hayes, R. (1991–2001) *National Retail Security Survey.* Gainesville, FL: University of Florida.

13 Hayes, R. (1997) Retail Crime Control: A New Operational Strategy. *Security Journal.* Vol. 8, No. 3, pp 225–32; Manunta, op cit.

14 Murphy and Criste, op cit.

15 Hayes, R. (2000) *NextLP: Protecting the Future Retail Enterprise.* Orlando, FL: Loss Prevention Research Council.

16 Ibid.

17 Murphy and Criste, op cit.

18 Hayes (2000) op cit.

19 DiLonardo, R.L. (1996) Defining and Measuring the Economic Benefit of Electronic Article Surveillance. *Security Journal.* Vol. 7, No. 1, pp 3–9; Hayes (2000) op cit.

20 Brainstorm 1995–1998. A series of two-day international conferences on loss prevention research and methods.

21 Ibid.

22 Bamfield and Hollinger, op cit.

23 Van Maanenberg, D. (1995) *Effective Retail Security: Protecting the Bottom Line.* Port Melbourne, Vic: Butterworth-Heinemann, p 191.

24 Hayes, R. (2001) *US Store Detectives: The Relationship Between Individual Characteristics and Job Performance.* Unpublished PhD thesis, University of Leicester.

25 Federal and Fogelman, op cit; Keckeisen, G. L. (1993) *Retail Security Versus the Shoplifter: Confronting the Shoplifter While Protecting the Merchant.* Springfield, IL: Charles C. Thomas, pp 26–30.

26 Hollinger, R.C. (1989) *Dishonesty in the Workplace.* Park Ridge, IL: London House.

27 Hayes, R. (1993) *Employee Theft Control.* Orlando, FL: Prevention Press; Hollinger, op cit.

28 Hayes (1991) op cit; Hayes (1993) op cit.

29 Hollinger *et al*, op cit.

30 Sherman, L.W., Gottfredson, D., MacKenzie, D., Eck, J., Reuter, P. and Bushway, S. (1997) *Preventing Crime: What Works, What Doesn't, What's Promising.* College Park, MD: University of Maryland and US Department of Justice.

Chapter 4

Tackling Shrinkage Throughout the Supply Chain: What Role for Automatic Product Identification?

Adrian Beck[1]

As the scale of the problem of shrinkage in the retail sector becomes more recognised, organisations are continually seeking news ways to try and tackle the problem. This chapter explores the possibilities presented by recent developments in the field of automatic product identification (Auto ID) and how it may be used to manage stock loss more effectively. It argues that while non-malicious forms of shrinkage such as process failures would be significantly reduced through the introduction of Auto ID, malicious shrinkage such as staff and customer theft presents more of a challenge. It concludes that Auto ID could play a powerful role in providing stock loss practitioners with unprecedented access to information about the movement of stock, which in turn could enable them to develop a more systemic and systematic approach to reducing stock loss throughout the supply chain.

Introduction

The purpose of this chapter is to explore the possibilities offered by recent developments within the sphere of product 'auto identification' (Auto ID) technologies to impact on the problem of shrinkage in the retail sector and of the losses incurred by the sector's suppliers (both manufacturers and third-party logistics providers). Its purpose is not to provide specific strategies for implementing Auto ID, but to consider the potential to impact upon shrinkage in the short, medium and long term. In this respect, this chapter aims to generate discussion, debate and above all raise questions about how this rapidly developing technology might be used, rather than to be overtly prescriptive in suggesting how it should be used.

It is based upon research carried out with European retailers and manufacturers, technology providers and developers, standards agencies and academics. The chapter begins by looking at recent developments in radio frequency identification (RFID) and tagging technologies, as well as the existing and, in some instances, well-established forms of product identification. It then goes on to chart the problems of shrinkage and the difficulties associated with tackling them. It continues with a review of some of the technological approaches currently adopted to tackle shrinkage, in particular the application of Electronic Article Surveillance (EAS). It then goes on to look at specific ways in which it might be used to tackle some of the problems outlined in the preceding sections. It concludes by assessing the overall issues relating to the implementation and eventual use of Auto ID in the business community.

Developments in product identification and RFID

Putting tags on items in order either to 'learn' something about them or to enable them to 'interact' in some way with other 'things' is nothing new. Electronic article surveillance tags have been around for about 40 years, and have been used primarily by the retail sector to try and counter the problem of shoplifting. Other systems have also been developed, most notably the barcode and the associated 'universal product code' (UPC), which have revolutionised the identification of products in the supply chain.[2] Through the Uniform Code Council (UCC) and the International Article Numbering Association (EAN International), barcodes have become a universally accepted standard, and also a ubiquitous part of product packaging. The UPC enables types of products to be identified optically, and for these products to be linked to databases that provide further information about them, primarily their price. UPC has had a major impact on the commercial world through the degree of global consensus it has required and achieved.

However, what the current barcode system does not do is identify each product uniquely—one bottle of shampoo is not distinguishable from another. It also relies on line-of-sight to gather the information, and therefore normally on human intervention to achieve this. These and other limitations have led to developments in RFID and the associated technologies relating to the means of carrying data (tags).

What is RFID?

Like a bar code, a radio frequency tag is a data carrier. While a bar code carries data in a visible symbol and is read at optical or infrared wavelengths, an RFID device (or tag) carries data programmed into a chip, and operates at various radio frequencies.[3]

All RFID systems have three main components:

- The RFID tag with its own data, functions and physical characteristics. Broadly speaking, all tags comprise a semi-conductor chip with memory processing capability and a transmitter connected to an antenna (aerial). The great majority of tags used within the retail environment are passive, taking their energy from the electromagnetic field emitted by readers.

- The reader, with its own functions and physical characteristics. This comprises an antenna, and a controller which codes, decodes, checks and stores the data, manages communications with the tags and communicates with the management system.

- The management system, with its own hardware and software. This is the nerve centre for the application, and forms part of the RFID user's information technology system. It is responsible for using the data received from and sent to the RFID tags for logistics and commercial management.

Figure 1. Radio frequency identification

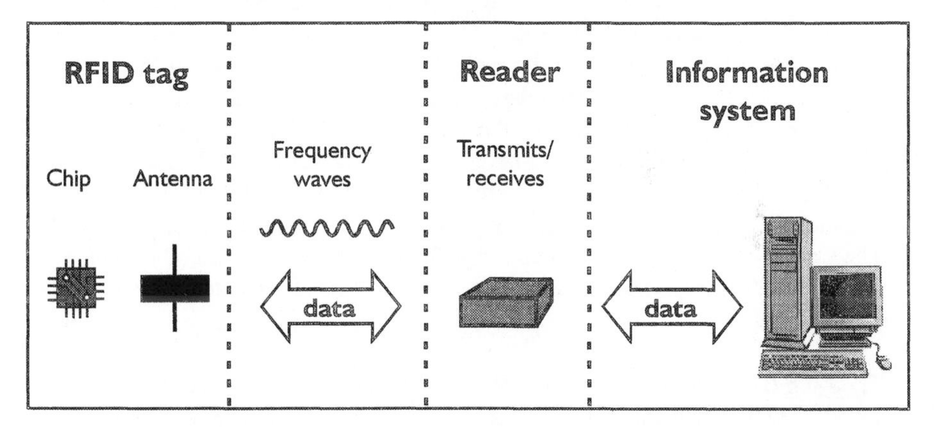

The perceived general benefits of RFID

As an RFID system uses radio waves, it can do things that optical technology cannot, including:

- *line of sight*: you do not have to be able to see a tag to be able to read it; it must merely pass through the electromagnetic field emitted by the reader;

- *range*: tags can be read at very long range—many hundreds of metres in the case of very specialised tags; RFID devices used in mass logistic applications are thought to need a range of around one to five metres;

- *bulk read*: many tags can be read in a short space of time; for example, the GTAG specification (see below) is designed to read 250 tags in less than three seconds;
- *selectivity*: potentially, specific data can be read from a tag;
- *durability*: tags can be protected from the elements or placed in a plastic casing;
- *read/write (RW)*: RFID tags can be updated after the original data has been loaded. This might be a simple change in status—'paid for' vs 'not paid for', in the case of a tag used in conjunction with retail EAS. It could, however, involve much more complex data, such as warranty and service history in the case of, for instance, a car or microwave oven.

Defining Auto ID

An emerging key player is the Auto-ID Center, which was founded at the Massachusetts Institute of Technology (MIT) in 1999. This is a global, industry-funded research programme tasked with developing the 'ultimate solution' for RFID: an ultra low-cost, open-standard system that can be used for any application on any physical object, including low-priced consumables such as grocery items. The Auto-ID Center mission is to 'merge the physical world with the information world to form a single, seamless network, using the latest technology developments embracing electro-magnetic identification, computer modelling and networking'.[4] The Center is working in collaboration with the global standards bodies and the Auto-ID industry to help specify how items will communicate in a standard environment.

The Auto ID Center is not without its critics. Some see the vision of the 'tagged world' as both unrealistic and unnecessary—the technological steps and the degree of collaboration required are formidable, and many of the problems it is designed to address can already be effectively tackled with existing technologies and better management systems.[5] There is no intention here of entering this debate. The purpose of this chapter is to envision a system where anything (pallets, cases or single items) have an RFID tag attached, and where these tags are capable of being effectively tracked along the entire supply chain[6] (through the use of widely distributed readers), and the resultant information is then capable of being analysed in real or near-real time. The term used here to describe this system will be 'Auto ID'. It is recognised that there are many potential benefits to organisations from this technology,[7] and that there may be considerable overlap between these benefits, but the emphasis of this chapter will be the impact of the technology only on the problems of shrinkage (see definition below).

The problem of shrinkage

The rather euphemistic term 'shrinkage' is used by the business world to describe the losses that occur while it attempts to complete the deceptively simple task of producing goods, distributing them, and eventually selling them to consumers. The term covers an enormous gamut of events, which can for the most part be broken down into two types: malicious and non-malicious. Malicious events represent those activities that are carried out to intentionally divest an organisation of goods, cash, services and, ultimately, profit. Non-malicious events, occurring within and between organisations, unintentionally cause loss, through poor processes, mistakes, bad design and so on. Both categories of events have a dramatic impact on the organisation's profitability.

The importance of perceiving the intentionality of a shrinkage occurrence is the impact on the approach adopted to tackle the problem. As the word implies, 'intentional' presumes deliberateness and a degree of forethought. It also presumes to a certain extent that existing systems have been found to be vulnerable—sometimes by accident, often by 'probing'—and in due course 'defeated' by the offender. Remedial action to deal with some types of malicious activity will therefore have a 'shelf life' or period of effectiveness that deteriorates as offenders find new ways to overcome such action. The latter could also lead to displacement—offenders target different products, locations, times or methods, or change the type of offence they commit.[8] On the other hand, unintentional shrinkage is usually less dynamic and more susceptible to lasting ameliorative actions. While it may require similar levels of vigilance (for instance, to make sure staff are continuing to follow procedures) it is less liable to be anything like as evolutionary in nature as its malicious counterpart.

Defining shrinkage

Opinions vary on a definition for shrinkage. Some take a very narrow perspective and limit it to the loss of stock only, choosing to exclude the loss of cash from an organisation, or consider it to relate only to the losses that cannot be explained—'unknown losses', as they are usually called.[9] At the other end of the spectrum, some argue for a much more inclusive, broad-ranging definition which encompasses both stock and cash, as well as the losses that result from shrinkage events—'indirect losses'—such as shop theft leading to products being out of stock,[10] the sale of stolen goods on the 'non-retail' market[11] or the production of counterfeit products. In addition, some feel that the expenditure incurred in responding to stock loss should also be included in the overall cost of shrinkage. This chapter will use the definition developed by Efficient Consumer Response (ECR) Europe's Shrinkage Group, based on four categories of shrinkage, encompassing both stock and cash, and made up of supplier fraud, internal theft, external theft and process failures. The first three can be regarded as malicious and intentional, while the fourth is non-malicious and an unintentional, but highly regrettable, consequence of ineffective business processes, procedures and activities.

The size of the problem

Recent research has once again demonstrated the extent of the problem of shrinkage for retailers and their suppliers throughout the world. In 2001, research sponsored by ECR Europe, on behalf of the 'fast moving consumer goods' sector (FMCG),[12] calculated that the annual bill for shrinkage was €18 billion, based on an annual turnover of €824.4 billion.[13] This equated to 2.31 per cent of turnover—1.75 per cent for retailers and 0.56 per cent for manufacturers.[14] A similar study in Australasia, using the same methodology, found that losses from shrinkage accounted for 1.73 per cent of turnover, and amounted to A$942 million.[15] In the USA, work by Hollinger has estimated that shrinkage costs the retail sector US$30 billion a year, or one-quarter of annual retail profits.[16] In some respects, there is nothing new about attempting to quantify the overall cost of stock loss[17] to the business world; the annual British Retail Consortium retail crime surveys provide a detailed breakdown of the extent and cost of the problem of crime against retailers in the UK, while similar initiatives in other European countries have also tried to measure the problem.[18]

While the definition of what constitutes shrinkage/stock loss varies between studies, and undoubtedly has an impact upon the overall size of the loss figure, the overriding conclusion is that the extent of the problem is enormous, and that it is an issue which seems, for the most part, extremely resistant to ameliorative actions.

Non-malicious shrinkage

Process failures. The 2001 ECR survey found that, for retailers, process failures accounted for 27 per cent of all losses, or €3.6 billion a year. For their suppliers, the percentage was much higher—78 per cent—but accounted for the same amount, €3.6 billion. Taken together, in Europe's FMCG sector process failures cost €7.2 billion a year, or €19 million a day. In the US, it has been estimated that for every $100 of shrinkage, $17.50 could be due to process failures.[19] This is a significant price to pay for organisations not getting it right.

The key elements that contribute to process failures ('paper shrink') are: stock going out of date; price reductions; damage to stock; delivery errors; pricing errors; scanning errors; incorrect inventory checks; product promotion errors; master file errors; and returns.

Common to most process failures is that they are a consequence of two related types of failure: a failure to *collect* and a failure to *communicate* information accurately and in a timely fashion about the products currently within the supply chain. There is a failure to answer two simple questions: 'what products do we have?' and 'where are they?' Answers to these questions then enable the key questions of 'what products do we need?', 'where do we need them to be?' and 'what price should we be charging for them?' to be answered.

Malicious shrinkage

Internal theft. The ECR survey estimated that for retailers 24 per cent of all losses, and for manufacturers 11 per cent, were due to internal theft, which accounts for just over €3.7 billion of loss each year. Despite this, companies, stock loss practitioners and indeed researchers have continued largely to ignore it as an area of concern,[20] choosing to focus more on the other problems affecting the sector, particularly external theft. Why has this tended to happen? There are four key reasons.

First, a lack of reliable, timely and detailed data on stock loss means that incidents of staff theft, compared with other forms of theft, are rarely recorded. Second, there has been a tendency for unknown losses to be apportioned to offenders outside the company. It is often more palatable to blame such losses on those who are not part of the store or company 'team'. For instance, unaccounted-for losses can be attributed to shoplifters—after all, research has shown that perhaps as few as one per cent of all shoplifters are ever caught.[21] Third, by its very nature staff theft can be a difficult crime to detect and investigate. Very often incidents of staff theft only come to light when a member of staff has stolen relatively large amounts of cash or produce. The difficulty in detecting and investigating staff crime leads to security managers opting to focus on other more accessible crimes such as shoplifting, burglaries, credit-card fraud and vandalism. Fourth, it has often been stated that security initiatives designed to target staff theft are bad for morale, that pointing the finger at staff can lead to a workforce that is poorly motivated and inefficient. Looking at the specific threats presented by staff, there are four key areas: theft of stock; 'grazing'; collusion, or 'sweethearting'; and theft of cash.

External theft. In stark contrast to internal theft, external theft has for the most part dominated the stock loss agenda. Despite numerous studies showing that it is not the single most significant threat to organisations,[22] it continues to receive the lion's share of shrinkage prevention expenditure. Partly this is because of some of the factors outlined in the previous section, but also because security service providers have played a significant part in setting the agenda and promising quick-fix technological panaceas. There are five main threats from external theft: shoplifting; returning stolen goods; 'grazing'; till snatches; and burglary.

Supplier fraud. The 2001 ECR Europe study of stock loss found that 12 per cent of all retailer losses were thought to be due to supplier fraud, which equates to

€1.6 billion a year. Supplier fraud is defined as the losses caused by suppliers, or their agents, deliberately delivering fewer goods than companies are eventually charged for. This includes vendor and contractor fraud, and the losses due to discrepancies in the goods supplied by third parties and not from companies' own distribution centres. The key forms of supplier fraud are: under/over delivery; phantom deliveries; deliberate invoice errors; returns; promotions; and discrepancies in the quality/weight of items delivered.

The critical aspect of many of the approaches adopted by suppliers to defraud the retailer at the point of delivery is that they exploit two key factors: the inability of most retailers accurately to check the delivery of items to a distribution centre or store, and the 'distance' between the point of delivery and the administrative/ordering function of the retail organisation. The sheer scale of deliveries to retail organisations means that it is almost impossible to check, certainly at item level, that what a supplier claims to have delivered has actually arrived. In addition, suppliers can exploit any disjunction between point of order and invoicing, and place of delivery—for example, when the buyer does not know that the products originally ordered have actually been delivered to the original specification, or when those responsible for billing are not fully informed about what has actually been delivered.

Responding to malicious shrinkage. As detailed earlier, the three types of malicious shrinkage outlined above, and the specific threats associated with each of them, present a significant challenge to stock loss practitioners. In particular, and of direct relevance to the use of Auto ID, is the evolutionary nature of the approaches adopted by offenders—for some of whom a new crime prevention strategy is perceived as less of a problem and more of a challenge. In addition, internal thieves are often in an ideal position to 'probe' new and existing processes and procedures to find loopholes that will enable them to increase their opportunities and reduce their risks. Recognising the organic nature of the approaches adopted to carry out malicious shrinkage is important in developing reduction strategies that are both realistic and responsive to this constantly changing offending environment. Those using Auto ID to tackle this type of shrinkage need to be fully aware of this.

Managing stock loss

Current problems

Responding to shrinkage has suffered from a number of inter-related problems that have combined to limit its effectiveness in dealing with an issue that is costing businesses billions of euros a year, in terms both of losses and of expenditure on so-called 'solutions'. Indeed, recent research has shown that if stock loss could be eliminated then the profits of a typical European retailer would be 58 per cent higher.[23] The factors undermining effective stock loss management are: its perceived peripheral status within organisations; a tendency to be uni-dimensional, reactionary and solution-driven; decision-making within an information vacuum; a lack of cross-functional organisational co-operation; and a poor appreciation, throughout the entire supply chain, of the threats posed.

Current shrinkage solutions

The types of approaches adopted by stock loss practitioners to tackle shrinkage can be broadly categorised into four main types: procedures and routines; people and processes: equipment and technology; and design and layout.[24]

The use of EAS within the category of equipment and technology is of particular interest to the debate on Auto ID, since the latter has been seen as a possible replacement/enhancement of EAS technology. This section, therefore, focuses particularly on this approach to stock loss reduction.

Electronic article surveillance. One of the chief methods adopted by many retailers to tackle shoplifting has been the use of EAS. These technologies have been in use for about 40 years, and are designed to increase the perceived risk of the offender being caught. The normal method is overtly marketing the presence of the system in the store. This is done in four ways: overt tags on products; the positioning of 'gates' at the entrances of stores; an audible alarm should the system be activated; and in-store notices alerting customers to the use of the system. The offender must then believe that the system will detect a tagged item leaving the store, that a member of staff will respond and apprehend the offender, and that the store will then proceed with some form of sanction (handing the offender over to the civil police for prosecution, and so on).

In theory this is an excellent form of crime prevention for dealing with the specific problem of shoplifting.[25] However, many difficulties have arisen, which have undermined the deterrent impact of these systems. One of the key problems has been the high level of false alarms (ie when the system is activated by a non-theft event). Some studies have found that as many as 93–96 per cent of activations are false alarms.[26] False activations of alarms can be caused by a wide range of factors, including: customers leaving the store with a tag that has not been properly deactivated by store staff; customers entering the store with a non-deactivated tag from another location; tags reactivating themselves after deactivation; and electrical items carried by customers triggering the system (bleepers, lap-top computers, etc[27]).

The impact of this has been markedly to reduce the confidence which store staff have in the system and to create a massive credibility gap (the 'crying wolf' syndrome[28]). This in turn has an impact upon the thought processes of the would-be offender: the likelihood of apprehension is perceived as much lower, and hence the rationale for offending is increased. In addition, professional shoplifters have become accustomed to finding ways of defeating the system. Once again, there are many methods adopted, including: removing the tag;[29] bending it; enclosing it within a substance that prevents it from sending a signal (shopping bags lined with aluminium foil, for instance); or purposely activating the system to enable others to leave with stolen goods while store staff are responding to (or resetting) the intentional system activation.

Current EAS systems also suffer from a lack of compatibility between competing proprietary technologies. There are currently four main types: acousto-magnetic (AM), radio frequency (RF), electromagnetic and microwave. Each has positive and negative aspects, depending upon the circumstances in which they are used.[30] However, the overall lack of standardisation and the considerable variation in sectoral and geographical adoption have created real problems for retailers and manufacturers alike. For instance, in the retail store environment some types of tags from one system can inadvertently trigger alarms in another system, while some products are unsuited to the application of tags (such as batteries or other metallic items). For manufacturers, meeting competing demands from numerous retailers to source tag products with a multitude of different types of EAS tag technologies[31] can add dramatically to production costs, not only through the initial cost of the tag but also because of the impact which applying the tag can have on rates of productivity.[32]

Evidence on the effect of EAS on levels of loss is mixed, with some studies suggesting it is very effective (particularly the use of 'hard' tags),[33] and others concluding it is of limited value.[34] Most of the studies, however, suffer from a lack of rigour in the way they have been carried out, undermining the extent to which lessons can be drawn from them. What would seem clear is that for certain types of offender (particularly the opportunist) the deterrent impact of tagging systems, if properly managed, is evident. The impact of EAS on the more determined offender is much less certain. Hence, the significant costs of installing and maintaining such systems are unlikely to be justified on the grounds of savings in reduced losses from theft. Above all, the limited scope of what the current EAS technologies are designed to achieve (they deal with shoplifting only, and simply notify staff when a tagged product has left the store without being deactivated)[35] should mean that their future role in stock loss management remains limited.

Meeting the challenges: what role for Auto ID?

If the technology providers and standards agencies can deliver the concept of a supply chain where each product can be uniquely identified, and tracked as it makes its way from the point of production through the distribution network and into the retail stores and beyond, then the potential benefits for stock loss management are dramatic and far-reaching. The purpose of this section of the chapter is hence to envision a supply chain where all objects can be identified and tracked automatically, and to consider how this may impact upon the problems of shrinkage and its management.

Process failures

Process failures are an area where item-level tagging could have a dramatic and profound effect. As detailed earlier, process failures are for the most part a non-malicious, unintentional outcome of a breakdown in the management of the movement of products through the supply chain. They are mainly caused by a lack of transparency in knowing where things are. The Auto ID scenario directly tackles this problem by providing a mechanism for tracking products automatically. The system will provide accurate and timely answers to the key questions: 'what products do we have?' and 'where are they?' By doing this, it will eliminate many of the problems outlined earlier. In particular, errors in inventory should become a thing of the past. Because product recognition and recording will be automated, the problems associated with accuracy (staff not counting stock properly) and timeliness (the physical and cost limitations of carrying out stock audits[36]) will be dramatically reduced. In effect, organisations would be able to maintain real-time inventories of their stock.

In addition, because staff will know exactly, and at any moment in time, what stock is currently out on the shelves, they will be able to manage the rotation and replenishment of stock better (and hence to reduce the amount that goes out of date or has to be reduced in price). Errors in delivery should also be reduced—receiving staff will know precisely what stock has arrived as it crosses the threshold of the building (which is particularly important for mixed pallet deliveries to stores). This information can then be automatically cross-checked with original order requests, and any discrepancies noted.

Auto ID should also markedly reduce the scanning errors made by till operators. Indeed, they could simply become product identification authenticators—providing a visual check between what has been read and what appears on the customer's receipt (this may be a reassuring middle ground between the current system and the future vision of a 'checkout-less' shopping environment)—and receivers of payment (not too dissimilar to the self-scan systems used by some supermarket chains).

Supplier fraud

Like process failures, supplier fraud could also be radically reduced (especially for retailers) through the introduction of Auto ID. The majority of supplier fraud occurs because the recipient of the goods is usually unable to physically check that the items claiming to be delivered have in fact arrived or are those that were ordered in the first place. Once again, the transparency and visibility of product provided by Auto ID is the key. Recipients of tagged stock will be able immediately, as it arrives, to cross-check the delivery note with the original order and with the goods presented at the point of delivery. Any overages and underages can be quickly identified and reconciled. Similarly, the authentication of each of the arriving products could be cross-checked with the manufacturer's product database to ensure the credibility of the delivered stock.

Internal theft

Because of the nature of this type of shrinkage, as outlined earlier, it is a difficult problem to solve, but Auto ID could play a role in a number of different ways.

Theft by staff. Most staff theft in retail stores is considered to take place in the back-room areas, where staff have the greatest opportunity for removing goods and remaining unobserved. They then have a number of options as to how the product is removed from the premises. There are four main ways: consume it; secrete it in their personal belongings and then carry it out as they leave; place it outside the store, usually in a garbage receptacle, for recovery at a later date; or use the internal post to have the items delivered to their home or some other 'safe' location.

The first of these methods is difficult to combat, and there is an argument that the normal physical limits on what staff can consume in a day preclude the need for such a problem to be prioritised. Carrying items out (either directly by staff or being dumped for later recovery) could be countered by tag readers being placed at all staff exists. Likewise, internal post boxes could either be located within known reader areas, or alternatively a reader could be attached to the post box itself. Either way, the existence of product within the post box will be capable of detection. The readers could then either activate an audible alarm and/or inform security staff of the illicit movement of goods outside the accepted boundaries. In this respect, it would act very much like the existing EAS tagging systems, but with the added benefit of security staff knowing the identity of the product that has triggered the alarm.[37]

Collusion. Staff colluding with shoplifters, particularly at the point of purchase, is notoriously difficult to detect. In the 'checkout-less' scenario,[38] the opportunity for this type of activity would be completely removed. The system would automatically 'detect' each and every product as it moved through the checkout area, nullifying any attempts either to avoid scanning the product in the first place or to 'fool' the system into thinking that one product was in fact another, cheaper item. Even if check-outs remained, the ability to 'mis-scan' a product intentionally would be dramatically reduced, as the item information would be transferred independently of the till operator.

Smart management: deterring dishonesty in the workplace. Perhaps one of the biggest impacts Auto ID may have on internal theft is through the deterrent impact of the information it can provide to managers and loss prevention staff. Internal theft is a function of opportunity and of a lack of disincentives. By radically improving the visibility of product and its movement throughout the working environment (including cases where it leaves through exits at the back of the building), not only should staff have less opportunity to remove goods, but also security managers should be able to respond more quickly and effectively to incidents as they occur. These two factors (supported by judicious advertising of the system by the company, reinforced by a tough policy for those caught stealing and by carefully planned access control) will act as a powerful deterrent to all staff.

External theft

Intelligent Electronic Article Surveillance (IEAS). As detailed earlier, current strategies for responding to shop theft have focused on the use of EAS technologies—usually based upon tags activating an audible alarm at the point of exit. This strategy requires a reaction from a member of staff, either to apprehend the offender or to confirm that goods have been legitimately purchased and that the tag has not been removed or deactivated. As research has shown, the level of false alarms on current EAS systems has resulted in extremely poor levels of confidence in the system by staff, and thus a low-level response to alarm activations at store exits. A dedicated shop thief knows this too, and to defeat the EAS will regularly rely on the apathy of store staff and their lack of belief in the system. For example, shoplifters working as a team will purposely trigger the system to enable others members of their team to then exit the store while the system is either deactivated or being reset by store staff.

One of the real dangers in the 'all item' tagged world, if the tags are used in the same way as current EAS tags (ie to trigger an alarm if the tag has not been deactivated at the point of sale), is that the level of reliability at the point of deactivation needs to be dramatically and consistently higher than currently achieved. If it is not, then the shopping mall of the future will resonate to a constant cacophony of alarms as a proportion of the many millions of tagged items moving from the point of sale to the store exit falsely trigger the security gates. This would not only irritate and embarrass honest customers, but also further reduce the confidence staff have in the system and provide yet more 'background noise' for shoplifters to exploit.

Dynamic hot product lists. What is perhaps more realistic is to be product-sensitive in the activation of the 'EAS component' of the tag. The notion of 'hot products'[39] is now a familiar concept within shrinkage management, where particular items are highlighted as being far more at risk of theft than others and hence deserve greater attention. In addition, by having access to better quality information on the products being targeted (through the greater visibility of products in the supply chain), stock loss managers could develop a much more dynamic, context-driven hot product list, enabling them to decide which products should be EAS-active given the local circumstances (such as store location).

Sweep thefts. Sweep thefts are a major problem for retailers, as they can result in significant losses through offenders taking large numbers of products from the shelf at one go. Developing a proactive response to this problem requires raising the risk of apprehension to the potential offender. This can be done in two interrelated ways. First, making the offender aware that the system has noted the number of products that have been removed from the shelf. This could be done through innovative active on-shelf displays, with remarks such as 'thank you for purchasing x number of x product' being shown. Second, through alerting in-store staff that a multiple number of pre-defined 'risky products' have been removed at once. They in turn could then be provided with additional information to enable them either to track the offender (for instance, through linking the activation to in-store CCTV), and or to be better prepared when the multiple stolen items eventually activate a store exit alarm. If this level of risk awareness and response could be achieved, it would send a powerful deterrent message to the offender community and further reduce the threat.

Tag visibility and deterrence. For the opportunistic shop thief at least, the deterrent impact of overt security measures has been found to be relatively successful. Overt security tags, and the presence of CCTV, security signs and security guards can all act to deter the casual offender. In order to maximise the security potential of a tagged product, the prospective offender needs to know that it has been tagged in the first place.[40] Much of the technological drive of Auto ID is to make the chips and antennae as small as possible, principally to enable them to be embedded in the product or the packaging, so that they are virtually invisible to the consumer. The danger with this is that any deterrent impact may be as small as the eventual chip! Therefore, careful consideration needs to be given to how the tags will be advertised. This can be done at three levels: at the micro level through notices on the product packaging; at the mezzo level through the use of displays in the shopping environment; and at the macro level by raising general public awareness through the media.

Removing tags. One of the problems with existing tagging systems is that the applied tag is vulnerable to removal by the committed and wily shop thief. Certainly some of the proposed ideas on embedding the tag within the product or the packaging will overcome many of the current tag removal strategies, but it may not necessarily stop those offenders who use other methods to nullify the ability of the chip to communicate with readers (such as placing the tagged product within an aluminium foil lined bag). The tracking capabilities of the proposed Auto ID system, however, may help with this problem too. For instance, the 'disappearance' of a tagged item between the shelf and the till could alert store staff to its possible theft. This would obviously, however, require a very dynamic and interactive information management system, together with an extremely prompt response from security staff, for it to be a realistic proposition.

Returning stolen goods. A common method adopted by shop thieves is to return stolen goods in order to try and get a cash refund. There are many variants of this, including: purchasing the same product as the one stolen, and then using the genuine receipt to refund the stolen item; using a stolen or invalid cheque book/credit card to purchase items, and then returning the goods for a cash refund; and intimidating store staff, claiming that receipt-less items were genuinely purchased. Whatever the method adopted, the thief relies on the same factor—that store staff are unable to tell whether the particular item being returned was ever purchased in the first place. By being able to identify each product uniquely, and know whether it has legitimately passed through a checkout, store staff will quickly be able to verify its status. In the situation where a stolen or invalid cheque book/credit card has been used, store staff should be able to link the item to the payment, and either return the cheque (if it is still within the store) or cancel the financial transaction. In turn, they should be able to better link the offender and the item through the transaction. Either way, the ability to identify individual products and their status (legitimately purchased or not) could be a powerful tool in reducing the losses created by refund fraud.

Developing an information-led strategic approach to shrinkage management

As detailed earlier, stock loss management is blighted by a number of key problems, not least the dearth of reliable data on how, when and where shrinkage occurs through the supply chain. Auto ID offers the very real prospect of providing shrinkage managers with a window on real stock loss. It could for the first time enable an accurate understanding of what percentage each of the shrinkage factors (internal and external theft, process failures and supplier fraud) actually make up of the whole. This would be a dramatic breakthrough, enabling stock loss managers to begin to develop a much more strategic approach to managing the problem.

By making the supply chain considerably more visible and transparent through the unique identification of all products, Auto ID could open up a whole series of opportunities for stock loss practitioners to make a significant impact on the current losses attributed to shrinkage. As detailed above, process failures and supplier fraud are particular areas that could be effectively targeted through the information made available by Auto ID. In Europe alone, this could lead to savings of €9 billion.

Within the areas of external and internal theft, the impact may be less dramatic, but there are certainly considerable benefits to be gained. With both problems, stock loss practitioners are disadvantaged by the delay between the incident occurring and its being detected (if at all). This time lag is particularly exploited by offenders within organisations, who quickly recognise the opportunities presented by complex, unwieldy and poorly managed stock inventories. Once again stock visibility data would provide a powerful lever for shrinkage managers to develop more effective and lasting processes and procedures to tackle the malicious theft of stock, at whatever point in the supply chain. This would come about not least because of the ability to create accurate and auditable records of accountability, increasingly connecting the movements of goods to people. Irrespective of the detection capabilities offered by this, the deterrent impact on staff could be significant, providing (if regularly reinforced) a truly proactive and lasting solution to the problem.

In terms of external theft, particularly shoplifting, simply considering Auto ID as a replacement for current EAS technology would be a mistake—the latter embodies a concept which is currently low on credibility. However, the opportunities provided by the information made available about the movement of products and their relationship with other products could enable stock loss practitioners to begin to target their resources much more effectively in dealing with this problem. It could certainly play a key role in relation to returns fraud and sweep thefts.

Indirect shrinkage

In the transparent supply chain, whereby stock loss managers can uniquely identify each product and provide an auditable trail of where it has been, they can begin to challenge some of the problems that have been previously perceived as beyond their remit/capabilities. The two key areas are counterfeiting and the sale of stolen goods in the 'non-retail' market. Counterfeiting of goods costs the EU business community €250 billion a year, while the figure for the global market is estimated to be $1000 billion.[41] In addition to this enormous loss of potential sales, counterfeiting can also have a detrimental impact on the reputation of a company, by inadvertently associating it with products that are sub-standard or dangerous. Manufacturers and their customers could, given the ability to identify all genuine products uniquely, quickly identify fake products entering the supply chain. Security managers could then collaborate with the policing agencies by providing auditable and evidence-quality records, and help in investigations to bring to account the organisations producing the counterfeit products.

Similarly, security managers could liaise with the public police to begin to address the sale of stolen goods. Car boot sales or flea markets have frequently been seen as an opportunity for recipients of stolen goods to sell them in the open market. Once again, the problem has been the inability to provide evidence that the goods on sale have been stolen (a similar situation to that of return frauds at retail stores, as detailed above). The police could be provided with readers that enabled them to gather data on the goods being sold. This information would then be linked directly to the manufacturer/retailer database, and the status of the items quickly established.

In both instances, the traceability of product is once again the key. But it also highlights the way in which the security manager of the future could take on a more expansive role, targeting problems that are seen as beyond their current remit but are malicious in nature and directly impact upon the bottom-line profitability of organisations.

Prospects, panaceas and practicalities

This chapter has deliberately avoided the debate about the feasibility and, indeed, desirability of introducing unique item-level automatic identification. There are considerable technological hurdles to be overcome before such a system could be introduced, and it is most unlikely that the Auto ID vision of all items being tagged will come about in the near future. However, Maoist determinism need not prevail, and a phased introduction is undoubtedly the way forward. Some products and some applications will benefit from earlier introduction than others, for example the tagging of selected products at pallet and case level. In terms of shrinkage management, this would clearly relate to those items perceived as hot products. Its introduction will however require common standards to be agreed, accepted and implemented.

In addition, the Auto ID vision is dependent on manufacturers and suppliers tagging their products at source, and ultimately paying for this in the first instance. The current debate on EAS source tagging is instructive on this issue. As pointed out earlier in this chapter, EAS tags are designed to deal only with the problem of shoplifting, and only really advantage those at the very end of the supply chain—they are uni-dimensional in purpose and in their prospective beneficiary. Quite rightly, manufacturers have difficulty in making a business case to support this strategy. However, the proposed Auto ID approach has many more potential benefits to offer the manufacturers, as it can be used to tackle directly some of *their* key shrinkage problems (for instance, in the FMCG sector in Europe, process failures account for 78 per cent of shrinkage or €3.6 billion annually, and this, as previously discussed, is one of the problem areas most likely to benefit dramatically from this approach). In other words, because Auto ID can be used to address a range of shrinkage issues throughout the supply chain, it is multi-dimensional in terms both of prospective beneficiaries and of problems addressed. Given this, the arguments against this type of source tagging become less persuasive.

It is important not to see Auto ID as a panacea (which it clearly is not) for the problems of shrinkage, but rather as a powerful *tool* to enable stock loss practitioners to manage the problem in a dramatically more effective way. Current stock loss prevention practice is characterised by a paucity of knowledge; this leads to responses which are piecemeal, partial, unsystematic and reactionary in nature, fuelled by those parts of the security sector committed to championing the use of proprietary technologies. Auto ID, by giving access to unparalleled levels of product information, could empower shrinkage managers to collaborate more successfully with the rest of their company and with other organisations across the supply chain, and to develop solutions that are effectively targeted, sustainable, and receptive to the constantly changing threats presented by shrinkage. In this respect, it fits neatly with much of the shrinkage reduction 'philosophy' developed by ECR Europe, including the use of the 'stock loss reduction road map' (see Chapter 12 in this volume) and the emphasis on developing approaches that are systematic, systemic and based upon inter- and intra-company collaboration throughout the supply chain[42]. Auto ID could play a pivotal role in enabling this approach to be both easier to adopt and more successful in achieving the desired outcome—reducing stock loss.

Notes

1 Adrian Beck is a Lecturer in the Scarman Centre, University of Leicester; email: bna@le.ac.uk.

2 Haberman, A. (ed.) (2000) *A Life Behind Bars*. Cambridge, MA: Harvard University Press.

3 Typically 125 KHz, 13.56 MHz, 2.45 GHz and around 900 MHz.

4 *http://www.autoidcenter.org/home_vision.asp*.

5 There are also many other concerns about the impact such a system might have on a range of other issues, not least the environment and the privacy of those who come into contact with tagged items.

6 In this context, the supply chain is considered to stretch from the point of production, through transportation and warehousing, to the retail outlet and reverse logistics (in the case of goods being returned). It also includes those items that enter the supply chain illegally (counterfeit) and leave it prematurely (for instance, for sale in the 'non-retail' market).

7 A whole host of potential uses have been identified, including: soft sensing of product freshness; faster product recalls; streamlined supply chains; improved compliance checking; interaction warnings; easier logistics for custom products; automatic recall; medication validation; etc.

8 Clarke, R.V. (1995) Opportunity-Reducing Crime Prevention Strategies and the Role of Motivation. In Wilkstrom, P., Clarke, R.V. and McCord, J. (eds) *Integrating Crime Prevention Strategies: Propensity and Opportunity*. Stockholm: National Council for Crime Prevention; Clarke, R.V. (ed.) (1997) *Situational Crime Prevention: Successful Case Studies*. Albany, NY: Harrow and Heston.

9 Masuda, B. (1992) Displacement vs Diffusion of Benefits and the Reduction of Inventory Losses in a Retail Environment. *Security Journal*. Vol. 3, No. 3, pp 131–6.

10 One study has suggested that between seven and ten per cent of product may be out of stock at any one time; see Efficient Consumer Response (Europe) (forthcoming) *Optimal Shelf Availability*. Brussels: ECR Europe.

11 Such as car boot sales or flea markets.

12 The term 'fast moving consumer goods' (FMCG) sector is used here to mean those retailers and their suppliers who provide a range of goods sold primarily through supermarkets and hypermarkets. The core of their businesses is providing 'essentials', such as various fresh and processed foodstuffs, but they also stock a wide selection of other goods, including health and beauty products, tobacco, alcohol, clothing, some electrical items, baby products and more general household items. Examples of FMCG retailers include Auchen, Carrefour, Coop Italia, ICA, Interspar, Tesco and Walmart. Examples of FMCG manufacturers include Allied Domecq, Gillette, Johnson and Johnson, Procter and Gamble, and Unilever. In the USA, this sector is also referred to as the 'consumer packaged goods sector'.

13 M+M Euro Trade (2000) *Trade Structures and the Top Retailers in the European Food Business*. Frankfurt: M+M Euro Trade.

14 Beck, A., Bilby, C. and Chapman, P. (2003) Shrinkage in Europe: Stock Loss in the Fast Moving Consumer Goods Sector. *Security Journal*. Vol. 16, No. 2, pp 61–75; Beck, A., Bilby, C., Chapman, P. and Harrison, A. (2001) *Shrinkage: Introducing a Collaborative Approach to Reducing Stock Loss in the Supply Chain*. Brussels: Efficient Consumer Response (Europe).

15 Phelps, D. (2002) *A Guide to Collaborative Loss Prevention*. Kingston, ACT: Efficient Consumer Response (Australasia).

16 Hollinger, R. and Hayes, R. (2000) *National Retail Security Survey*. Gainesville, FL: University of Florida.

17 The terms 'shrinkage' and 'stock loss' will be used interchangeably throughout this chapter.

18 For the UK, see the British Retail Consortium's annual reports on the costs of crime to the retail sector; see also Mirrlees-Black, C. and Ross, A. (1995) *Crime Against Retail and Manufacturing Premises: Findings from the 1994 Commercial Victimisation Survey*. Research Study No. 146. London: Home Office. For Germany, see EuroHandelsinstitut (2000) *Inventurdifferenzen 2000: Ergebnisse einer aktuellen Erhebung*. Cologne: EuroHandelsinstitut; for data (on theft only) from European retailers, see Bamfield, J. (2002) *The European Retail Theft Barometer*. Nottingham: Centre for Retail Research, and Chapter 8 of this book.

19 Hollinger and Hayes, op cit.

20 See Mars, G. (1982) *Cheats at Work: Anthropology of Workplace Crime*. London: Unwin; Hollinger, R., Greenberg, J. and Scott, K (1996) Why Do Workers Bite the Hands That Feed Them? Employee Theft as a Social Exchange Process.

Research into Organisational Behaviour. Vol. 1, pp 111–56; Beck, A. and Willis, A. (1993) Employee Theft: A Profile of Staff Dishonesty in the Retail Sector. *Journal of Asset Protection & Financial Crime*. Vol. 1, May, pp 45–57; Beck, A. and Willis, A. (1995) *An Evaluation of Security Hardware*. Vol. 4. Leicester: University of Leicester; Bamfield, J. (1998) A Breach of Trust: Employee Collusion and Theft from Major Retailers. In Gill, M. (ed.) *Crime at Work: Increasing the Risk for Offenders*. Leicester: Perpetuity Press.

21 See Arboleda-Florez, J., Durie, H. and Costello, J. (1977) Shoplifting: An Ordinary Crime? *International Journal of Offender Therapy and Comparative Criminology*. Vol. 21, No. 3, pp 201–7; Murphy, D. (1986) *Customers and Thieves*. Aldershot: Gower.

22 Buckle, A. and Farrington, D. (1984) An Observational Study of Shoplifting. *British Journal of Criminology*. Vol. 24, No. 1, pp 63–73.

23 This is based on research conducted by Cranfield School of Management.

24 See Beck *et al* (2001) op cit.

25 It should be noted that EAS technologies are not designed to deal with many of the other problems faced by retailers, such as burglary, robbery, internal theft, process failures or supplier fraud.

26 Handford, M. (1994) Electronic Tagging in Action: A Case Study in Retailing. In Gill, M. (ed.) *Crime at Work: Studies in Security and Crime Prevention*. Leicester: Perpetuity Press; Beck and Willis (1995) op cit.

27 The objects that can activate different types of EAS system are almost legion, ranging from pacemakers and metal prostheses to personal identity cards and library books.

28 Handford, op cit; Shapland J. (1995) Preventing Retail Sector Crimes. In Tonry, M. and Farrington, D. (eds) *Building a Safer Society: Strategic Approaches to Crime Prevention*. Chicago, IL: University of Chicago Press.

29 Some retailers have argued that source tagging would reduce this problem, enabling manufacturers more effectively to incorporate the tag into the product or its packaging. While this may be true, there are considerable difficulties in implementing EAS source tagging, not least the lack of agreement on a standard global system (what type of tag should manufacturers adopt, will it have to vary around the world, and will it have to vary depending upon the retailer being supplied?). Such matters can have a dramatic effect on the manufacturer's production costs, with companies such as Gillette estimating that applying EAS tags to their products would slow down their production lines by as much as 35 per cent. In addition, the question of who pays for the tag is a thorny issue, particularly given that the sole purpose of EAS is to deal only with shoplifting. Not surprisingly, manufacturers are reluctant to invest in a technology that has a dubious track record and will deliver little real value to their businesses. (This issue is discussed again at the end of this chapter.)

30 There appears to be little consensus within the EAS industry about which technology is most suitable for use in the retail environment, although there has

been some recent discussion on creating a single industry-standard security tag for the US market. See IDTechEx Limited (2002) EAS Defeated in the US. *Smart Label Analyst*. No. 15, pp 23–4.

31 Some companies have resorted to applying more than one type of tag to a product.

32 Methods to deal with the impact on production rates have included 'fractionalisation', whereby goods are selectively tagged at the point of production (for instance, every third item). The deterrent aspect of this approach obviously relies on the tags being very clearly 'advertised' to would-be offenders.

33 Some recent data from the US on the impact of EAS source tagging of pre-recorded music and video products suggests that it has been effective in reducing losses, although it did take 20 years to reach a consensus on what product should be used. See Wanke, E. (2002) How EAS Source Tagging Rewrote Shrinkage History in the Music and Video Sector. *Loss Prevention*, May-June, pp 23–5.

34 Sherman, L., Gottfredson, D., Mackenzie, D., Eck, J., Reuter, P. and Bushway, S. (1997) *Preventing Crime: What Works, What Doesn't, What's Promising*. Washington, DC: Office of Justice Programs, US Department of Justice.

35 The author recognises that some EAS management systems are now slightly more sophisticated, and can interact with EPOS systems and record what has caused tag activations. However, their scope is still considered to be relatively limited.

36 Hayes, R. (1991) *Retail Security and Loss Prevention*. Stoneham, MA: Butterworth-Heinemann; Jones, P. (1998) *Retail Loss Control*. Oxford: Butterworth.

37 This works on the assumption that staff have not been able to remove the tag. If the stolen products were for future re-sale, then overly damaging the packaging (in order to remove or deactivate the tag) would be counter-productive. However, many of the problems, highlighted earlier, concerning professional shop thieves and their attempts to counter EAS tags may also apply to members of staff engaged in stealing from the company.

38 One of the elements of the 'vision' put forward by the Auto ID Center at MIT is that in the future stores may not need tills, as customers would be billed automatically once the products had passed through the store exit.

39 Clarke R.V. (1999) *Hot Products: Understanding, Anticipating and Reducing Demand for Stolen Goods*. Police Research Series, No. 112. London: Home Office.

40 A study by Beck and Willis found that only 13.4 per cent of customers were aware of the presence of EAS tags (see Beck and Willis (1995) op cit), while Hayes and Rogers found that 86 per cent of apprehended shoplifters were unaware that products were EAS-tagged (quoted in Hayes, R. (2002) *EAS Impact Analysis: A White Paper*. Unpublished).

41 *http://www.aacp.org.uk/cost-intro.html.*

42 Beck *et al* (2001) op cit.

Chapter 5

Organising and Controlling Payment Card Fraud: Fraudsters and Their Operational Environment

Michael Levi[1]

This chapter discusses how payment card fraud has changed and how it is organised in relation to the prevention efforts of the card business and of retailers. Various techniques have had a significant impact on the ease with which genuine cards can be obtained and used, and recent rises in counterfeits and Internet/mail-order frauds look disproportionate because these other areas are better controlled. Variations in social and intellectual capital affect crime and its organisation. To the extent that 'plastic' fraudsters can remain relatively unobtrusive, unconnected with 'heavy' crimes such as drugs and robbery, there is no need for them to 'get organised' in order to enjoy a profitable career: it is easy and quick for British criminals to operate in different regions of the country, or even on the Continent. The implications of smart cards plus PIN are explored, along with other key areas of public-private partnership policing.

Introduction

This chapter starts on the assumption that there have been and will continue to be people who are willing to commit serious crimes for gain, to whatever extent they are allowed to get way with and until they deem that they are running risks they are not prepared to tolerate. In that rather trivial sense, the 'rational offender' model is appropriate, but it does not tell us much. In addition to any personality and age dimension which may affect the appetite for risk-taking—and such appetite may be greater among offenders and senior business executives than among the general public or security personnel—the key problem is to relate criminal decision-making to the context in which individuals find themselves or (and this requires more motivation) are able to place themselves after actively seeking out opportunities. The shape of their relationships depends on how many of them are willing to engage in particular crimes (or, for some, several types of crime); on what structures of criminal association already exist in their 'communities' or can be created by them; and on how 'society'—including but not restricted to the commercial sector—is organised to combat the crimes in question. Before considering the question of 'organised crime' involvement in any given crime, we should evaluate what that crime *requires* in the way of organisation and skill sets.[2] A growing number of studies demonstrate the variability of these preconditions for criminal work[3] and the present author has argued elsewhere that involvement with traditional 'organised crime groups' may not be optimal for professional criminals, especially not for business criminals who thereby receive more police attention than they would otherwise get.[4]

This chapter examines the interaction between crime and its regulation in the context of 'plastic fraud', a term which includes bank-issued and retail store credit cards, debit cards, charge cards (such as classic American Express, which have to be paid at the end of the month), and cheque guarantee cards. This differs in some key features from non-fraud types of crime for gain. Obviously many 'hot products'[5] can be used globally, and some types of crime may not even require physical movement of stolen products—cell phone frauds, for example (though very few criminal users end up in cells). World payment systems—Visa, MasterCard/Europay and American Express—have been created and exist to allow people to pay for a product anywhere on the globe without requiring the vendor to make any assessment of them, other than whether their signature roughly matches that on the back of the card. Vendors pay a fee—itself related (a) to the risk estimated to be posed by them (since they may be fraudsters), (b) to the general fraud and bad debt levels in the particular market, and (c) to general competition for business—for the guarantee of payment and for the extra business they receive by virtue of offering the payment card facility. (Countries and businesses that do not offer credit card payments will have few American or other Western business visitors. To minimise downside exposure for bankers with 'risky clients' and/or in 'risky countries', some cards cannot be used internationally and have pre-set spending limits.) This is a vast, complex and highly differentiated market, containing many commercial actors with competing interests.

The extent of the criminal market for plastic fraud

The continual battle between the industry and criminals has made the past decade something of a roller-coaster ride for the payment card industry. At the beginning of the 1990s, plastic card fraud stood at around £165 million per annum, after a dramatic rise in the late 1980s. The first Home Office-commissioned study on the prevention of cheque and credit card fraud[6] was followed by a raft of measures and a fall to £83 million in 1995, despite much higher turnover and many more cards in circulation. Levi and Handley[7] estimated that without the fraud prevention measures introduced during the period 1991–96, the 1996 fraud costs would have been 350 per cent higher. However, in the seven years from 1995 fraud losses on payment cards rose annually, approaching £400 million in 2001 (see Figure 1.) The police recorded 153,281 cheque and payment card frauds in the 12 months to April 2002,[8] but the card companies themselves refer few cases to them, since there is little prospect of action against which to offset the issuer or acquirer staff time and cost involved, and reporting 'dud' cases undermines the credibility of the commercial victims and their representatives the next time they want to see some action taken.

Figure 1. Gross card fraud rates in the UK

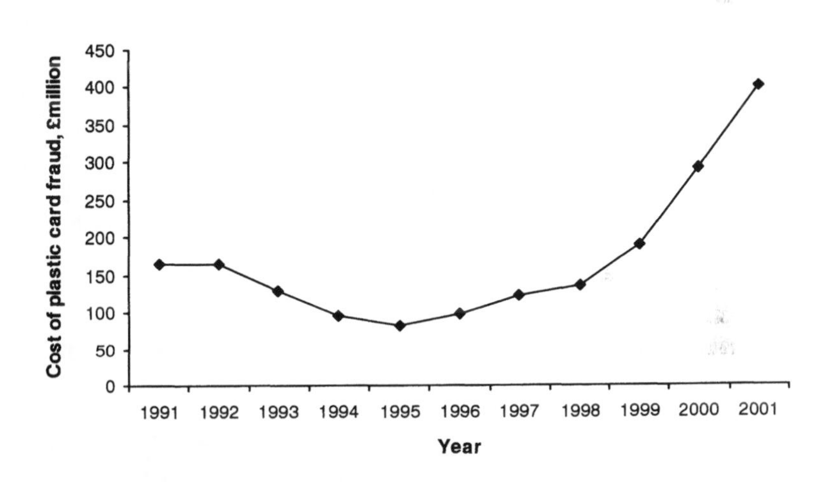

Within this overall rise, there has been a significant change in the *pattern* of fraud. The £114 million of fraud on lost and stolen cards (which until 2000 constituted the largest proportion of loss) stands at 80 per cent of the 1991 level; so too does the £26.7 million of fraud on cards diverted before delivery to the customer (which had dropped to half its 1991 level until a rise in 2001). By contrast, fraud on counterfeit cards has now surpassed that on lost and stolen cards and, at £160.3 million in 2001, stood at more than 22 times the 1991 level. The £95.7 million of fraudulent transactions where the card is not present—for example telephone, mail-order and Internet purchases—have shown a startling rise (albeit from initially low levels) to around 150 times the 1991 level (though some of these losses fall on retailers). (See Table 1.) Both counterfeiting and card-not-present (CNP) fraud further split off fraud opportunities from the necessity to commit a traditional property crime.

Table 1. Fraud losses on UK-issued cards split by fraud type

	£million		
	2000	**2001**	**Change (%)**
Counterfeit cards	107.1	160.3	50
Cards stolen or lost	101.9	114.0	12
Card-not-present (mail, telephone, Internet)	72.9	95.7	31
Cards intercepted in the post	17.7	26.7	51
Fraudulent applications	10.5	6.6	-37
Other	6.9	8.0	15
Total	**317.0**	**411.4**	**30**
Fraud losses against turnover	0.162	0.183	13

Finally, a growing area of 'populist' concern has been identity theft and fraud,[9] which cost an estimated £12 million in 2001, made up of £6.6 million for application fraud and an estimated £5 million for account take-over. Application fraud occurs when criminals use stolen or false documents—like genuine or counterfeited utility bills and bank statements—to build up useable information to open an account. With 'account takeover', criminals first gather information about the intended victim, then contact the card issuer to request that mail be redirected, and later request a replacement for a lost card.

Barriers and opportunities in the commission of plastic crime

The debate about the involvement of organised crime in payment card fraud, or indeed in other types of crime, risks oversimplification and over-homogenisation. It is oversimplified because it assumes that payment card fraud represents a single phenomenon rather than a set of sometimes quite discrete activities: the discovery of credit cards during a burglary or theft from a car and their resale or use in stores requires a different skill set and criminal environment than the copying of magnetic stripe data, transmission overseas and the manufacture of a counterfeit card for use around the world. It is over-homogenised because it trades upon our images of 'organised crime' drawn from *The Godfather* or media portrayals of a Chinese and Russian 'red menace'. There is evidence of Russian, triad and Italian Mafia involvement in plastic fraud—including counterfeiting factories, and using a reputation for violence to pressurise merchants into processing phoney or counterfeit card transactions through their stores—and there is evidence of payment card fraud being used to fund the Tamil Tigers and some Islamic terrorist groups; but there is no evidence that payment card fraud generally is organised on hierarchical lines or funds terrorism (though if other sources of terrorist finance dry up, it may attain a greater significance in future). Indeed, even where many people are recruited and go out and use counterfeit cards or card data at the same time, there is no *need* for participants to be members of long-term groups. Inertia plays a part in how criminals behave, provided that their opportunities are not reduced or eliminated: Mativat and Tremblay[10] suggest that Quebecois fraudsters were reluctant to shift from using stolen cards to altering cards (despite the ready availability of cheap and compact encoding and embossing equipment), although there may have been some point on the ease and reward axes at which they would have made the change. The present author would argue that variations in social and intellectual capital will affect life chances in the criminal just as in the licit world.

The most difficult form of payment card fraud to get substantial funds from is ATM fraud, which (other than payment when physically present at the bank counter) is the only one that generates cash direct to the offender. Many cases of ATM fraud occur when legitimate cardholders have written down their PIN—to avoid forgetting it—and kept it with their card in a purse or wallet that is then stolen. An increasing number of cases involve shoulder surfing, where criminals look over a cash machine user's shoulder (in more than one scheme, from a vantage point using telescopic sights) to watch them enter their PIN and then steal the card using distraction techniques, pick pocketing or street robbery. Also said by industry sources to be on the rise are the use of devices that trap the card inside the ATM, at which point the criminal approaches the victim and tricks them into re-entering the PIN. After the cardholder gives up and leaves, the criminal removes the device and the card and withdraws cash. However, these work best for offenders when they are rare enough for cardholders to be unaware of the risk. Fraud at ATMs in the UK cost the banking industry £21.2 million in 2001, representing 5.2 per cent of total fraud losses. Other than these relatively rare plastic frauds where cash is obtained, fraudsters need to sell the goods that they obtain.

To illuminate some factors that bear upon criminal decision-making and organisation, the present author and some associates[11] conducted semi-structured interviews with 28 fraudsters in various parts of England and Wales, found by 'snowball' sampling: these were located in London, the Midlands, the North West and Wales. Although the representativeness of an unknowable population always gives (or should give) rise to concern, there is no reason to suspect that these interviewees present a distorted portrait of opportunity structures for or constraints on fraud, though obviously it would have been better to have had more interviewees, and one might have wished for a wider *range*—especially of international fraudsters, none of whom were interviewed—as well as enhanced representativeness.

Six in-depth interviews were conducted over several sessions; the other interviews were more superficial. In addition, the present author interviewed several heads of police 'cheque squads' from around the UK about fraudsters' lifestyles. Typically, plastic and cheque fraudsters' personal mobility tends to be within 'their' region, varying from a city to a 50-mile radius (using the motorway network to facilitate rapid turnover). That said, plastic fraudsters' mobility is greater than for most offenders, who generally commit offences within a couple of miles of their homes. Between a quarter and a third of fraud against UK-issued cards occurs overseas, especially in France and Spain, though interviews suggest that this is due principally to foreign thieves and fraudsters rather than to British migrants operating overseas. There are also major counterfeiting rings, often involving overseas-origin British residents and temporary workers who copy magnetic stripe and other personal data and transmit it overseas to Hong Kong, Malaysia, Sri Lanka or Russia, for use on cards to purchase goods for resale from various parts of the world. The key operational stages for offenders are reviewed below. The 'primary offenders' (eg burglars, robbers, hoteliers, waiters and merchants) may not move, but the product does and their collaborators do, sometimes transnationally.

Stage 1. Financing plastic fraud

Few forms of plastic fraud require the reconnaissance time and 'business expenses' of sophisticated robberies or burglaries, drugs trafficking or longer-term fraudulent schemes. Nor, except for either the physical counterfeiting of cards and holograms or the establishment of phoney merchants with the principal aim of defrauding the card companies, are there the sort of initial capital costs associated with bulk drugs and/or automatic weapons purchases: travel, accommodation and dining can be paid for with the cards themselves, unless this creates some extra risk of exposure. (Many offenders pay for flights with stolen cards which, if the police took a more active interest, would be very risky, since they would be tied to a particular location within a particular time frame.)

Stage 2. Obtaining cards or card-like financial instruments

How do offenders obtain cards or card-like instruments?

- Fraudulent credit applications, sometimes using data and documents such as bank statements and utility bills (obtained from bin-raiding) that can be used as 'proof' of identity or address. A survey for Experian[12] discovered that three-quarters of UK councils were aware that bins were raided for data in their area, of which four-fifths said it had got worse in recent years. Furthermore, one in five bins (more in wealthier areas) contained a whole credit or debit card number—mostly including an expiry date—and bank account details that could be used for fraudulent purposes.

- Genuine cards obtained from theft, burglary and robbery—the British Crime Survey shows a fluctuating but rising trend in the proportion of burglaries in which credit cards or cheque books have been stolen, from seven per cent in 1993 to 11 per cent in 2001–02, peaking at 15 per cent in 1997. In addition there is simple loss, plus the theft of unsigned cards in transit.

- Manufacturing counterfeit cards that would deceive most ordinary merchants, using numbers copied from genuine cardholders or generated by algorithm programmes obtained from the Internet.

- Cruder counterfeit transactions committed with the consent of the merchant, from shop assistant to store manager.

- Telephone and mail-order purchases using genuine card numbers, with goods delivered to addresses other than those of the true cardholders.

Applications frauds. These can be committed by individuals or by groups. The stereotype involves people of Nigerian origin—whether living permanently in the UK or merely visiting—who use personal data obtained by cleaners, and/or 'empty house' address data obtained from contacts in council housing departments, to develop 'non-suspicious' applications for cards. If they are risk-averse and have multiple mail-drops for cards available (perhaps through corrupt local council officials), some applications fraudsters use names and addresses only once or twice, to reduce the chance that they will be spotted by the industry's sophisticated database systems, which search for any discrepancies in personal circumstances on different credit applications, as well as for frauds associated with the same name and address. Unless the applicant is merely 'borrowing' the identity and creditworthiness of a real person, they may need 'front' companies or conspirators with genuine businesses to supply them with their false employment and income references.

Stolen driving licences or those obtained in false names, plus stolen or computer-generated fictitious utilities bills, can be used to open accounts, particularly favouring 'instant credit' places where credit controllers may have to make decisions within at most five minutes. (In the long run, however, when existing licences are replaced, the introduction of photographic driving licence cards should make this difficult, without expensive counterfeiting.) To allay suspicions, offenders might claim that they had only just moved to where they were now, and therefore were not on the electoral register yet. Offenders learn from other fraudsters and through media reports when prevention efforts are stepped up in response to losses.[13]

Manufacture of counterfeit cards. Syndicates with hi-tech abilities and financial backing manufacture cards with plastic, holograms and embossing of variable quality. They are usually Chinese-origin gangs based in the Far East, where covert plants are established.[14] The cards are then distributed through gang members in their global network, including Canada as well as Europe.[15] The gangs obtain the card numbers which are required to 'pass' the system checks, either from genuine cards that they 'clone' after obtaining details from collusive retail outlets (an aspect of retailer collusion or even initiation overlooked by Mativat and Tremblay), or from down-loadable Internet number generators which tell the world at large how to generate logically possible numbers in the 'bin ranges' that card issuers use, which do not correspond to real numbers in issue and therefore are not 'clones'. The latter are simply another manifestation of the range of criminal possibilities to which the Internet gives rise.[16] 'Buying in' expertise from corrupt electronic specialists ('project crime'—see McIntosh[17]), more adventurous criminals have attempted such things as tapping the line between ATM and bank—though the transmitted data are encrypted and include a Message Authentication Code which changes with every transaction—thus far, according to industry sources, without success.

An alternative to careful counterfeit manufacture is to produce 'white plastic', ie cards embossed with genuine card-holder details, obtained by taking an extra copy of those details, or by loading electronic data from point-of-sale tills into a laptop. The industry discourages retailer collusion by charging back to the retailer transactions demonstrably fictitious and/or by the risk of being placed on a 'warning' list operated by Visa UK on behalf of the industry as a whole. However, there is usually a 'window of opportunity' before such intervention occurs.

Thefts of cards. The best cards to steal are those without signatures on them taken before the intended cardholder realises they are missing: hence the attractiveness to postal workers or sneak thieves of cards in transit. Targeted special card deliveries in high-risk areas cut such losses from £32.9 m. in 1991 to £12.5 million in 1997, before rising again recently for reasons not fully understood.

The next most desirable cards are high-value ones whose owners will not immediately notice their absence. Sometimes the more professional walk-in hotel and office thieves just steal one card from a wallet, reasonably assuming that the person will not notice or take immediate action, for they know it will normally take some time before the victim notices that his/her card is missing and deduces that it has been stolen. Hotel, club-house, and night-club/bar security is obviously relevant to this. Bar-tenders can also make use of cards left with them by groups of drinkers who are having a lengthy bar 'tab'.

In London, one in five street robbers obtain cards from their victims,[18] but, though many of these cards are used, sophisticates in the interview sample avoided cards that came from aggravated burglaries or muggings, not so much out of morality but because these have priority police interest, and they are worried that 'intelligence-led' policing will try to solve them through the credit card transactions; their object (not always realised because of ill-discipline) is to stay low-key and outside police targeting. Cards from muggings supposedly go to 'the blacks', who (allegedly) do not care and need them to pay for their drugs.

The cards from a variety of sources having been 'used up' by the risk-averse professionals, some are passed on in the secondary market for approximately £20, reflecting the greater risk and lower potential yield before exhaustion of credit. Such 'second phase' cards are used to buy groceries, drinks, cigarettes, etc.

Stage 3: Obtaining goods or funds from cards and disposing of them
The proportion of lost and stolen cards subsequently misused is surprisingly low—about one in 12—reflecting both that some cards are simply lost around the house and also the varied 'efficiency' levels of criminal markets; many thieves (perhaps especially juveniles[19]) do not have contacts to buy their cards, so throw them away; and others lack the 'bottle' for upfront self-presentation using the cards. Except where burglars know that householders are on holiday and therefore are unlikely to have reported the cards as stolen, the electronic transmission to retailers' point-of-sale terminals of 'hot card' files makes it rational for offenders to spend heavily and quickly. It also displaces *some* card usage to the few outlets (such as off-licences) which do not have modern electronic point-of-sale terminals.

There has been some controversy over the issue of retailer vigilance. An unknown percentage of recovered cards are clawed back only when the electronic system instructs staff to do so rather than by proactive vigilance. There is also variation in whether the store keeps for itself rewards for staff of £50 (rising to £500 for 'skimmed' cards in which the magnetic stripe details differ from the names and details on the face of the card). Research suggests that staff vigilance over signatures seldom rises above the moderate[20] and offenders interviewed were well aware of this; there is both a 'bottle' and a subjective risk assessment component in *repeated* use of cards that may have been reported stolen. One way for fraudsters to negotiate environmental risk is to have friends in retail outlets who can tell them about those control measures of which they have knowledge, and/or to use stores where employee awareness is known to be low.

Except for direct use (for example cigarettes and alcohol), most goods obtained by fraud, like stolen goods generally, are resold, probably in car boot sales, pubs and housing estates.[21] Interviews suggest that as some thieves in poor estates become older, they become 'general wheelers and dealers' (including drugs) and both buy cards from the 'younger generation' of burglars and send them or others in their network out on 'shopping expeditions' with the stolen cards. Some diversify into other areas, such as social security fraud, insurance fraud and other forms of credit fraud which require similar sorts of skills.

What factors affect the *scale* of fraud and which *particular* outlets are targeted? Interviews suggest that fraudsters have 'comfort zones' such as stores or parts of stores where they feel more or less confident. Some believe that they can turn the *shopkeepers'* anxiety to their own advantage by making *them* feel guilty. Few believe that being challenged will lead to arrest, unless they are particularly careless (which includes being too 'stoned' to function properly). One skilful fraudster stated that experience and technical knowledge enable him to understand the rhythm of a normal transaction, and to know precisely what people are likely to be saying and when. So if something is out of place—for example, if the merchant seems to take an abnormally long time to process a transaction—things are about to 'come on top' and he leaves (if necessary with his fingerprints on the card in the retailer's possession). Technology reduces individual suspiciousness and speeds up the process of buying, while retailer phone calls increase the time of exposure, and fraudsters like to get in and out quickly. If he wants an expensive video-camera from a multiple retailer, he will produce a plausible rationale, eg he has a friend who has one or saw a demo of it last week. This would allay any suspicion as to why someone is buying such an expensive item in a couple of minutes without seeing how good it is and how it works. (Cynically, he assumes that if the salespeople and/or firms are paid by the merchant acquirer, they will not care whether or not the card is genuinely his—what the present author terms the 'rational blind-eye' of narrow corporate self-interest.) There is always some adrenalin in the fear of forgetting the name on the current credit card, particularly in a restaurant or hotel: this has to be covered up, for instance with 'sorry—absent-minded professor'!

Card issuers' literature often tells holders never to divulge their PIN even to people claiming to be from the bank, but cardholders can sometimes be conned into divulging to people claiming to be from the issuer who 'need it for security reasons'. In any event, this is a low-risk option unless they are phoning from home and have not blocked their home number!

Some interviewees learned what to do by acting as 'look-out' for a while before 'fronting' the fraud. To be skilled at 'card present' payment card fraud does require 'bottle' but does not require much sophistication or the high interpersonal front skills that might be required for advance fee frauds, still less for major commercial frauds. Fraudsters in strange geographical environments might use contacts they have met in prison or through drugs networks, but the routines of large retail chains are much the same anywhere.

The majority interviewed were predominantly 'criminal opportunists' who drifted in and out of fraud. Those more heavily into an 'underworld' are permanently on the lookout for gaps in security systems, such as allowing customers to order further cheque books long before they have exhausted the others. Some scan commercial brochures and websites to see if they can spot a criminal opportunity, including variations in practices between different banks and building societies; others pick things up from other prisoners and/or from watching television programmes/reading newspaper articles about card fraud and prevention measures.

One offender stated that he and his friends do not normally steal (ie defraud) to order, but rather mention to their local bartender that they have a particular item for sale, and the bartender asks around for potential buyers.

Offenders' objectives are also affected by how ambitious they are and who they meet. Opportunities for women are quite good in this area of crime. Sometimes, there are travelling teams who go around in cars, trawling regional shopping centres; the effects of CCTV on this are unknown, but—where images are good enough—CCTV offers a potential data source to be mined for prevention and criminal justice intervention if the data are stored. The more risk-averse offenders in travelling teams use the cards only for one day before throwing them away or re-selling them: the supply of cards is plentiful enough to make this feasible.

It is not argued here that payment card fraud requires no skill or preparation: these are necessary, especially in the longer term when luck can run out. Informed fraudsters try to check whether stolen cards are on a 'hot card file' by test purchases in low-risk environments, and make purchases just under what they estimate or are told are the authorisation limits for their cards; thus systems that remotely sample below the normal limits can be quite effective in picking such transactions up. Fraudsters do run risks at the point of sale (though staff concerns about violence may give fraudsters greater optimism, especially when the store itself is unlikely to lose money). However, police operations, such as surveillance, and action against merchants who collude with fraudsters to pass genuine or counterfeit cards in large numbers, may be difficult without the provision of substantial private-sector assistance by the card schemes and, sometimes, by the Association for Payments Clearing Services and individual card issuers. This, especially when dealing with international and national 'skimming' and counterfeiting networks, is the rationale for the establishment in 2002 of a Dedicated Cheque and Plastic Crime Unit, funded by the industry (three-quarters) and Home Office (one-quarter).

The reduction of plastic fraud: concluding comments

There have been very substantial changes in payment card prevention since the early 1990s, which have sought to reduce opportunities for those forms of fraud with the most alarming rate of increase or absolute levels.[22] Perhaps the biggest strategic move was the widespread introduction of smart cards to replace (eventually) the magnetic stripe, due to the awareness that the latter was inherently highly vulnerable to easy skimming and to the counterfeiting of cards that duplicated the details on existing ones. The ways in which fraudster groups are able to operate depends largely upon their technical and organisational battles with card issuers, merchant transaction processors and retailers; with the exception of CNP transactions, retailers suffer least from fraud (or suffer only indirectly, via their merchant service charges), and therefore are least motivated to sacrifice speed of customer service in order to reduce fraud. One consequence has been resistance to the shift to a chip card with PIN since 2001, since retailers expect very few benefits, and larger retailers consider that changes will cost them money and trouble in changing their in-house electronic systems. Chip cards with PIN are not invulnerable to attack in the longer term, but in the absence of a pure biometric which authenticates the cardholder as well as the card (and which can check card-and-cardholder-not-physically-present situations as well), they offer a much tougher obstacle to easy, volume use of counterfeit and stolen cards than does the present set of technologies.

If and when chip cards with PIN are implemented, they will displace some fraud internationally to jurisdictions which do not have smart cards, via more developed crime networks (and international travellers such as truck drivers); and the UK will remain attractive for fraudsters using foreign cards (such as American ones, which are unlikely to become 'smart' until much later) which may retain magnetic stripes, and therefore cannot be protected using this technology. The industry must plan on the assumption that at some stage encryption will be broken and compromised, and that application frauds (which have not risen as much as expected) and merchant frauds will become more popular. The growth of remote Internet sales will enable new fake businesses to be created which pretend to be authorised to accept credit card payments but actually exist only to capture card data (and the 'Card-holder Verification' digits printed on the back of contemporary cards which do not appear on the magnetic stripe), for later use in counterfeits; but card companies will monitor the Internet to try to close them down quickly. Faster liaison between firms and the willingness of the police to act quickly on information will continue to be crucial in reducing the zone of free experimentation that currently exists for fraudulent attempts. However, given the different pace and economic cost-benefit balance of fraud prevention activities in different regions of the globe, and the necessity for cards to have global utility, a rapid-response system that picks up fraud trends early and looks at local as well as international interventions looks like the best bet for continuing to reduce payment card fraud.

Notes

1 Michael Levi is Professor of Criminology, Cardiff University; email: Levi@Cardiff.ac.uk.

2 Levi, M. (2002) The Organisation of Serious Crimes. In Maguire, M., Morgan, R. and Reiner, R. (eds) *The Oxford Handbook of Criminology*. 3rd edn. Oxford: Oxford University Press.

3 See, for example, Gill, M. (2001) The Craft of Robbers of Cash-in-transit Vans: Crime Facilitators and the Entrepreneurial Approach. *International Journal of the Sociology of Law*. Vol. 29, No. 3, pp 277–91.

4 Levi, M. (1981) *The Phantom Capitalists: The Organisation and Control of Long-Firm Fraud.* London: Heinemann; Levi, M. and Naylor, T. (2000) *Organised Crime, the Organisation of Crime, and the Organisation of Business.* London: Crime Foresight Panel, Department of Trade and Industry.

5 Clarke, R. (1999) *Hot Products: Understanding, Anticipating and Reducing Demand for Stolen Goods.* London: Home Office.

6 Levi, M., Bissell, P. and Richardson, T. (1991) *The Prevention of Cheque and Credit Card Fraud.* Crime Prevention Unit Paper No. 26. London: Home Office.

7 Levi, M. and Handley, J. (1998) *The Prevention of Plastic and Cheque Fraud Revisited.* Home Office Research Study No. 182. London: Home Office.

8 Simmons, J. (2002) *Crime in England and Wales 2001–2002.* London: Home Office.

9 Performance and Innovation Unit (2002) *Privacy and Data Sharing.* London: Cabinet Office. It is noted here (without giving a great deal of underlying reasoning) that: 'It is difficult to calculate the cost of identity fraud to the UK economy, but the available evidence suggests that it is at least £1.2 billion each year – although this is almost certainly an underestimate' (para. 3.29).

10 Mativat, F. and Tremblay, P. (1997) Counterfeiting Credit Cards. *British Journal of Criminology.* Vol. 37, No. 2, pp 165–83.

11 The author is grateful to Karen Evans and Audrey Stephenson-Burton for their help in the offender interviews, and to Mary Bosworth for her help in analysing police data.

12 Experian (2002) *An Experian White Paper: Lifting the Lid off Identity Theft and Transaction Fraud.* Nottingham: Experian.

13 Though this is assuming a degree of self-control which, though prized and asserted in theory, is often not found in practice among offenders, who observation suggests are usually far less rational and disciplined than they believe themselves to be or claim to be—they have idealised self-images. Moreover, even if they are in control of their own activity levels, they would normally have little control over others practising the same techniques. The more such individuals there are (through the spreading of information), the faster card issuers' tolerance levels will be breached.

14 Newton, J. (1994) *Organised Plastic Counterfeiting.* Police Research Group. London: Home Office.

15 Ibid; Mativat and Tremblay, op cit; author's interviews.

16 Levi, M. and Pithouse, A. (forthcoming) *White-Collar Crime and its Victims: the Media and Social Construction of Business Fraud.* Oxford: Clarendon; Mann, D. and Sutton, M. (1998) NetCrime: More Change in the Organisation of Thieving. *British Journal of Criminology.* Vol. 38, No. 2, pp 201–29; Grabosky, P. and Smith, R. (1998) *Crime in the Digital Age.* Brunswick, NJ: Transaction; Grabosky, P., Smith, R. and Dempsey, G. (2001) *Electronic Theft: Unlawful Acquisition in Cyberspace.* Cambridge: Cambridge University Press.

17 McIntosh, M. (1975) *The Organisation of Crime.* London: Macmillan.

18 Pease, K. (1997) personal communication.

19 Sutton, M. (1998) *Handling Stolen Goods and Theft: A Market Reduction Approach.* Home Office Research Study No. 178. London: Home Office.

20 Levi and Handley, op cit.

21 Sutton, M., Schneider, J. and Hetherington, S. (2001) *Tackling Theft with the Market Reduction Approach.* Crime Reduction Paper No. 8. London: Home Office; Sutton, op cit.

22 Levi, M. and Handley, J. (2002) *Criminal Justice and the Future of Payment Card Fraud.* London: Institute for Public Policy Research.

Chapter 6

Can Information Technology Help in the Search for Money Laundering? The Views of Financial Companies

Martin Gill and Geoff Taylor[1]

The fight against money laundering has taken a variety of forms. One crucial element is the production by finance companies of 'Suspicious Transaction Reports', which provide information for investigation agencies. Clearly, the greater the grounds of suspicion—or more specifically, the more evidence there is to support a belief that a transaction is illegal—the better the base for enforcement action. Yet very little is known about how finance companies generate intelligence about suspicion from the billions of transaction that pass through accounts each day. This chapter considers this issue, reports on finance companies' experiences and views, and specifically assesses the role of IT in this process. The findings suggest that in tackling money laundering IT does not yet play a major role, and many, including users of dedicated systems, are sceptical about its value. The chapter concludes that this may reflect more a lack of understanding about 'suspicion' than a weakness of IT systems per se.

Introduction

To prevent financial crime in general, and money laundering in particular, financial companies are required to observe 'due diligence' with respect to their customers by establishing 'know your customer' (KYC) procedures. These include the need, discussed elsewhere (Gill and Taylor, forthcoming), properly to identify customers, and the need to monitor account usage with a view to detecting suspicious transactions. Indeed, there is a regulatory requirement to identify suspicious transactions and to report them in the form of a 'suspicious transaction report' (STR) to the National Criminal Intelligence Service (NCIS), who in turn pass on reports to relevant authorities. These latter will often be Financial Intelligence Units within local police forces, but if the report involves tax evasion it will be passed to HM Customs or the Inland Revenue as appropriate. The simple idea here is that financial institutions identify suspicious activities, and thus guide enforcement authorities in arresting and prosecuting money launderers. In short, the role of the STR is central to the fight against money laundering.

The Basel Committee on Banking Supervision[2] articulated the importance of this role in the fight against money laundering:

> Ongoing monitoring is an essential aspect of effective KYC procedures. Banks can only effectively control and reduce their [money laundering] risk if they have an understanding of normal and reasonable account activity of their customers so that they have a means of identifying transactions which fall outside the regular pattern of an account's activity. Without such knowledge, they are likely to fail in their duty to report suspicious transactions to the appropriate authorities in cases where they are required to do so.[3]

But what precisely is meant by 'suspicion', and how can it best be identified? The JMLSG[4] recognises that identifying suspicion involves a degree of subjectivity; however, it argues that suspicion is more than 'mere speculation' and has some supporting base which may fall short of actual proof:

> A degree of satisfaction not necessarily amounting to belief at least extending beyond speculation as to whether an event has occurred or not.[5]

Not surprisingly the JMLSG takes care to spell out the process by which as much information as possible is made available to a legally responsible individual, the Money Laundering Reporting Officer (MLRO), to make a decision. So while there is a legal requirement to report suspicions, much discretion is left to companies, and in particular to MLROs, about how this should be done. The Financial Services Authority sourcebook is also rather vague on this issue, merely requiring that firms 'take reasonable care to establish and maintain appropriate systems' (para 7.2.1).[6] Nevertheless, there have been calls for more use of automated computer software.[7]

The very diversity of the financial sector, and the size and account portfolios of different financial companies, mean that these aims are easier to achieve in some companies than others. Some financial institutions, for example private banking, have a highly personalised style of working and may be in a position to task account managers with providing the appropriate levels of oversight. For other institutions, such as high-street clearing banks or those with national and even international reach through a branch network, the requirement to be familiar with customers and all their account behaviours is far from straightforward. In such circumstances there is a tendency to look to technology and software to monitor transactions to provide the necessary level of oversight. Certainly there are those who have strongly advocated IT as a positive step in producing better STRs.[8]

The aim of this chapter is to discuss findings of a study involving a survey of 1337 financial companies to elicit their views on different aspects of regulation. A synopsis of the findings is presented elsewhere.[9] In this chapter we examine financial companies' views and experiences in monitoring transactions to tackle money laundering, more specifically using IT to generate information for consideration as STRs. The chapter begins by reviewing the process of monitoring transactions, in particular with an assessment of the criteria used by financial institutions to determine suspicion. The chapter then reports on the finance sector's experience of using software to detect money laundering, either by making use of a system that is already in place, sometimes in an adapted form, or by employing dedicated software. It will be shown that while practices vary, there is widespread scepticism about the value of IT in detecting suspicious activity.[10] The chapter ends with a discussion of the implications of these findings.

Monitoring transactions and determining suspicion

Locating suspicious activity through monitoring transactions can be broken down into three stages. The first involves *monitoring accounts* to gain understanding of routine account behaviour. The second is the *exceptional activity* stage, and involves establishing the criteria to be used to separate routine from exceptional or unusual activity. Finally, the *proving suspicious* stage involves the scrutiny of exceptional activities to determine whether it should be reported to NCIS. In practice the stages run together—and indeed the meanings of words to describe them are problematic[11]—but there are two common approaches used to begin the process of determining if account activity is unusual. One approach is to set a fixed parameter and examine any transaction which meets or exceeds it, for example by setting a transaction figure. This approach is used in Australia and the US; in the latter the legislative cornerstone to counter money laundering is the 1970 Bank Secrecy Act, which requires financial institutions to file details of every transaction over $10,000 in the form of a 'Currency Transaction Report', a system which has the merit of a clear reporting requirement but which has proved very expensive:

> The banking industry calculates the cost at approximately $130 million per year, or between $3 and $15 per CTR report. In addition it costs the federal agencies approximately $2 per report to process and store the data.[12]

The other approach, and the one adopted in the UK, requires financial institutions to notify the regulatory authorities of any circumstances sufficiently out of the ordinary to cause suspicion. While cheaper than the American system, the cost lies in the loss of clarity about when to report:

> The only flaws with Suspicious Activity Reports is that a degree of discretion is left in the hands of clerks, tellers, compliance officers, and other financial institution personnel as to whether the transaction or transactions are suspicious; what is suspicious to one may not be suspicious to another, and a crooked teller or clerk can easily avoid noticing anything 'suspicious'.[13]

Indeed, some have questioned whether this is even practical:

> ... it seems to me that it is unreasonable to expect legitimate financial institutions to be able to distinguish between unusual transactions including tax evasion or drug trafficking. From the banks' standpoint they simply see an amount of money which is unusual which forms the basis of suspicion.[14]

Certainly, there is a lack of clarity about the criteria used by financial institutions to monitor transactions in determining the activity which needs to be reviewed for 'suspicion' (ie the second, *exceptional activity* stage) and so the study sought to address this topic. Specifically, respondents were presented with a range of criteria which could be used to identify suspicious activity and were asked to state whether these were used, and also whether the criteria were considered in isolation or alongside other factors. The findings are shown in Table 1.

Table 1. Criteria used to identify potentially suspicious activity

Evaluating internal activity	Criteria considered in isolation	Criteria considered in conjunction
Criteria such as:		
transactions over a particular sum, n = 333[15]	18.6% (62)	45.9% (153)
transactions by certain clients, n = 333	9.0% (30)	46.5% (155)
transactions from certain locations, n = 333	6.3% (21)	48.0% (160)
transactions from offshore entities, n = 331	5.4% (18)	42.9% (142)
transactions which vary from previous client profiles, n = 332	11.1% (37)	59.0% (196)
Transactions which vary from normal account, product or service trading behaviour, n = 331	12.1% (40)	61.6% (204)

A clear preference emerged for taking all issues into account when considering suspicious activity, but respondents placed greatest emphasis on variations from normal customer or account behaviour. Just one parameter stood out as worthy of merit in its own right, and that was when transactions breached a pre-determined ceiling, the system first used in the US.

Monitoring accounts

As indicated above, account monitoring is easier in some companies than in others; smaller firms with account managers and a personalised service are better placed to monitor activity than larger institutions, with many customers involved in thousands if not millions of separate transactions every day, many of which are processed automatically. Moreover, the validity of particular criteria will vary because different business types will have alternative perceptions of what is normal and routine and what amounts to unusual activity. Other influences are changes to clients' circumstances, which add to the problem of classifying what is normal. Taking personal accounts as an example, routine behaviour can be influenced by developments such as promotion or job change with potentially more disposable income, or redundancy with the opposite effect; people may receive an inheritance or win the lottery or take up new hobbies, all of which could explain sudden fluctuations in an account profile, as Levi notes:

> How many bankers have any idea what their customers do, particularly those commercial customers that do not need to borrow money, because they are merging the cash from crime into their businesses?[16]

Hence the complexity of the task in monitoring accounts and the importance of systems which can help to identify 'unusual' activity that may also prove to be suspicious.

The process is costly, and since this is after all a matter of business it is not perhaps surprising in the modern age that some companies should turn to technology for assistance. The main attraction of software is that computers can process much larger quantities of data more reliably than humans. However, the costs involve capital outlay for the necessary program (which can be expensive) and development costs, including the need to generate links to the in-house system. There may well be a period of heightened risk while the new system 'beds in', and areas of incompatibility are discovered and managed. Furthermore there will be staff time involved, plus a training programme. The benefits are better account monitoring and improved chances of detecting those using the company for the purposes of money laundering; and there is a potential for big savings in that there are enormous costs associated with aiding money laundering, including action from the FSA for failing to comply with regulations and the costs of damage to reputation.

The findings set out in Figure 1 show that just under half the sample (46.9 per cent) used software to monitor transactions, and that only a small minority (6.2 per cent) had dedicated anti-money-laundering software.

Figure 1. Use of software for the purpose of monitoring transactions

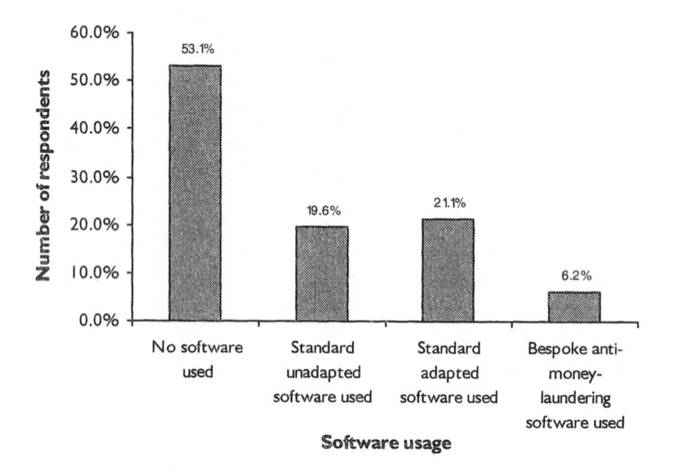

The use of dedicated anti-money-laundering software was very rare (just 21 respondents), most software users employing in-house systems to monitor transactions. It was found that using IT to monitor transactions was more common in banks and building societies compared with investment and insurance companies, as shown in Figure 2 below.

Figure 2. The use of software to monitor transactions, by different business types

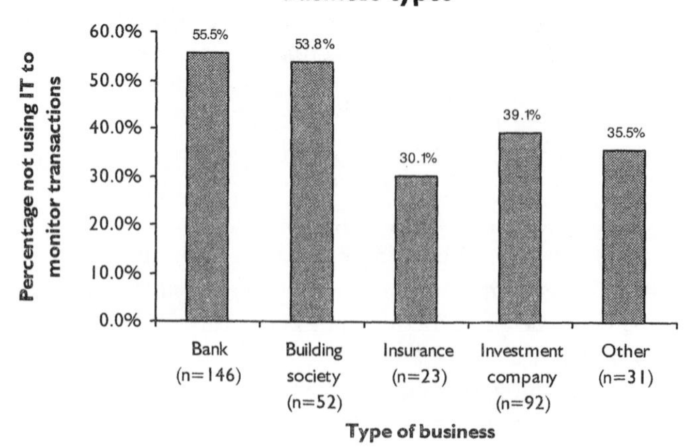

While banks were marginally the biggest users of software to monitor transactions, further analysis established that retail banks (64.1 per cent, n = 39) were the heaviest users of IT to monitor transactions, closely followed by corporate banks (52.2 per cent, n = 23). Given the logic of the argument linking the number of clients and the need for increased processing power to monitor transactions, analysis was conducted of the relationship between company size and the different methods of transaction monitoring identified by the survey. Due to the resource-intensive nature of the dedicated programs, it might be anticipated that they would be more common in larger companies; our findings confirmed this.

Table 2. Use of IT to monitor transactions, by size of institution

	Smaller institutions (up to 1000 staff)	**Larger institutions (over 1000 staff)**
IT used to monitor transactions	55.9% (147)	43.6% (34)
IT not used to monitor transactions	44.1% (116)	56.4% (44)
N =	**263**	**78**

One reason why companies may have elected not to use dedicated anti-money-laundering software, aside from cost, was the potential to use in-house systems to monitor transactions, either in the existing *unadapted* form, or *adapted* for the purpose of controlling money laundering. Once again findings varied by company size, as shown in Table 3.

Table 3. Respondents using in-house software to monitor transactions

Form of software used to monitor transactions	**Smaller institutions (up to 1000 staff)**	**Larger institutions (over 1000 staff)**
Unadapted in-house system	51.0% (52)	40.5% (15)
Adapted in-house system	49.0% (50)	59.5% (22)
N =	**102**	**37**

Respondents monitoring transactions manually

A full picture of the respondents monitoring transactions manually, ie not using software, is provided in Table 4.

These findings confirm that most smaller institutions relied on manual processing to monitor transactions (possibly indicating the lack of need for processing power in providing the necessary level of oversight), and that not every large institution used software for this purpose.

Table 4. Respondents relying on manual processing to monitor transactions, by size and business type

	Banks	Building societies	Insurance companies	Investment companies
Smaller institutions (up to 1000 staff)	84.6% (55)	100% (24)	62.5% (10)	76.8% (43)
Larger institutions (over 1000 staff)	15.4% (10)	-	37.5% (6)	23.2% (13)
N =	**65**	**24**	**16**	**56**

To summarise, smaller institutions were less inclined to use software than larger firms, although results differed by type of institution. Among those which used IT to monitor transactions, as many used the in-house system in its original form as adapted it; institutions only rarely acquired dedicated programs. The next section explores why this might be the case.

Concerns about dedicated anti-money-laundering software systems

Respondents were asked for their views about the suitability of dedicated software. Their views on four key points were examined: the extent to which they felt they could rely on these systems to monitor transactions; the extent to which they accepted the principle of using automated systems to tackle money laundering; whether reliance on software was appropriate; and costs. The findings are presented in Figure 3.

At least two interesting issues emerged here. First, many respondents did not attach a high priority to the use of software. Well over half the sample believed that transaction monitoring could be achieved without dedicated software. Only a fifth disagreed with the contention that there was too much reliance on software to counter money laundering, and many (about half those answering the question) neither agreed nor disagreed. Moreover, nearly half felt that dedicated software was not needed to carry out transaction monitoring properly. Second, the jury is still out on whether software reduces the cost of identifying suspicious transactions. While over a third offered neither a positive or negative view, there were about as many arguing that it reduced costs as there were arguing the opposite.

Figure 3. Respondents' views on the effectiveness of bespoke software to monitor transactions

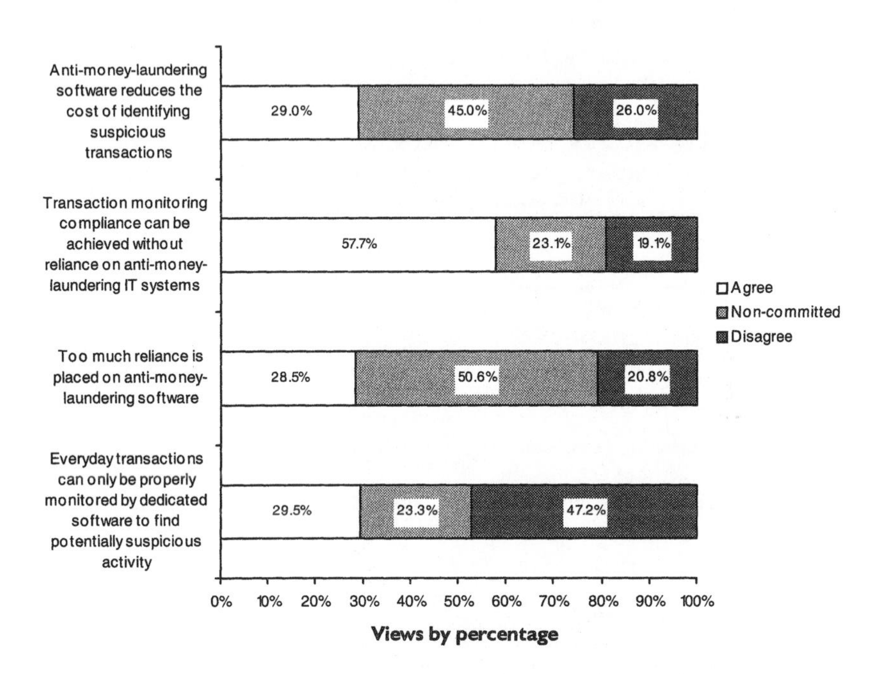

However, it was possible to take the analysis a stage further and investigate these views according to whether the respondent used software to monitor transactions. These findings, which do not distinguish between different levels of software usage (ie between adapted and unadapted in-house systems, or dedicated programs), are presented in Table 5.

While around three-quarters of the sample remained negative about the need for IT to monitor transactions, regardless of whether they used it, the question on cost effectiveness generated a difference. The majority of respondents who did not use IT felt it would reduce costs, whereas only the minority held this view amongst those who presently used IT. There was perhaps some naïve optimism here! Clearly, the relatively negative perceptions about anti-money-laundering software were not confined to those not using software to monitor transactions.

There is one other finding of relevance here (discussed more fully elsewhere[17]), that those who monitored transactions using IT made more disclosures to NCIS.[18] So while we know that large companies use IT more and produce more reports we do not know whether they are better-quality reports, ie that they provide more evidence and therefore more leads to money launderers.[19] The use of expensive software may or may not be improving money-laundering detection, but the question remains whether the key driver for institutions in justifying the cost of such software is to detect money laundering, or to avoid the regulator's displeasure and perhaps to provide an 'insurance policy' in the event that their services are in fact found to be being used to launder funds.

Table 5. Respondents' views on dedicated anti-money-laundering software based on their own use of IT to monitor transactions

Respondents' views	Yes	No
Anti-money-laundering software reduces the cost of identifying suspicious transactions:		
Respondents using IT, n = 70	47.1% (33)	52.9% (37)
Respondents not using IT, n = 53	60.3% (32)	39.7% (21)
Transaction monitoring compliance can be achieved without reliance on anti-money-laundering IT systems:		
Respondents using IT, n = 111	73.9% (82)	26.1% (29)
Respondents not using IT, n = 123	76.4% (94)	23.6% (29)
Too much reliance is placed on anti-money-laundering software:		
Respondents using IT, n = 70	57.1% (40)	42.9% (30)
Respondents not using IT, n = 56	58.9% (33)	41.1% (23)
Dedicated software is needed properly to monitor transactions:		
Respondents using IT, n = 115	44.3% (51)	55.7% (64)
Respondents not using IT, n = 130	33.1% (43)	66.9% (87)

Conclusion

We need to put these findings in perspective. It was found that just under half of the sample resorted to software programs to achieve a satisfactory level of monitoring of everyday business, but only a few had fully embraced the software route by installing dedicated anti-money-laundering programs. Examination of the different business types showed that banks and building societies were the main users of IT to monitor transactions, a practice less likely in other business types. The survey was able to correlate company size (and therefore likely size of client base) with the adoption of IT to monitor transactions, although some larger institutions made no use of software for this function.

Despite the adoption of software in this role among almost half of the sample, the majority view of dedicated anti-money-laundering software was not positive. This view was not influenced by the respondent's own use of automated systems, which may indicate a degree of dissatisfaction with the performance of such systems in this task. The majority view was that software was not needed and that the process could be reliably completed without dedicated anti-money-laundering software. The only issues which divided respondents was cost-effectiveness. Marginally less than half the IT users, but a clear majority of non-users, thought dedicated systems were cost-effective.

The lack of overt endorsement of anti-money-laundering software may have several causes, and some may not have been considered within the survey. Nevertheless, an overriding finding was that most respondents believed the process of finding unusual activity was a complex one requiring consideration of a number of issues, and the negative attitude to software may stem from a lack of confidence that any software could reliably perform this function. There was also a view that systems need a human intervention at some point in the chain, and so it was simply not possible to rely exclusively on software, as one respondent noted:

> In large companies IT systems may weed out transactions but cost time and money to develop and install. In smaller companies the huge expensive IT systems cannot be justified given the volume [of transactions]. But whatever method is used it relies on someone viewing the items and making a decision.

Evaluating the effectiveness of technology in spotting suspicious activity is very difficult when it is not clear whether the reports lead to money laundering. It is too early to suggest that technology is faulty, and/or that people are not using it wisely or to best effect, although both these suggestions may be true. It seems that at least one major problem is the lack of a clear understanding about the elements of suspicion that later lead to correctly defining a money-laundering transaction, to generating intelligence about future offenders, or to making future offences more difficult. Here the responsibility rests not with those who design IT systems but with the FSA, NCIS and the police. Until this issue is tackled there remains the impression of a lot of wasted (and very expensive) effort, comprising potentially unproductive investment in technology and increased costs (no doubt passed on to customers). Meanwhile, effort may well be being diverted away from drug and people smugglers, terrorists and serious and serial offenders.

Notes

1 Martin Gill is Director, Perpetuity Research and Consultancy International; email: m.gill@perpetuitygroup.com. Geoff Taylor is Community Safety Co-ordinator, Birmingham City Council. The authors would like to thank Andersens, who sponsored this study, and in particular Mike Adlem, Nic Carrington, Paul Doxey and Deepak Haria for their constructive advice on all aspects of the project, Karryn Loveday for help with the research and John Walker for comments on a draft of this chapter.

2 Known as the Basel Committee, this consists of senior banking regulators from ten countries—Belgium, Canada, France, Germany, Italy, Japan, Luxembourg, the Netherlands, Spain, Sweden, Switzerland, the United Kingdom and the United States. Its role is to formulate broad standards and guidelines, and to recommend statements of best practice in the area of banking supervision

3 Basel Committee on Banking Supervision (2001) *Customer Due Diligence for Banks* Basel: Bank for International Settlements, p 13.

4 The 'Joint Money Laundering Steering Group' is an industry-based consultation group which meets under the auspices of the British Bankers' Association; see *www.bba.org.uk/docs/JMLSG.*

5 JMLSG Guidance Notes 2001, para. 5.3.

6 Cynically, one might argue that if the FSA advocated a specific type of system then it would have to address the costs, and therefore refrains from any specific comment on this issue.

7 Performance and Innovations Unit (2002) *Recovering the Proceeds of Crime.* London: Cabinet Office. For a good discussion, see *Money Laundering Bulletin.* No 79, pp 9–11.

8 See *Money Laundering Report.* Nos. 79–80.

9 Gill, M. and Taylor, G. (forthcoming a) *Preventing Money Laundering Or Obstructing Business? Financial Companies' Perspectives On 'Know Your Customer' Procedures*; Gill, M. and Taylor, G. (forthcoming b), *Policing Money Laundering: Understanding the Disparity in Reporting Suspicious Activity Amongst the UK Financial Sector.*

10 The FSA pledge to carry out a study 'to identify ways in which it may be able to use IT to improve regulatory processes further and to reduce direct and indirect regulatory costs' has yet to materialise. (Financial Services Authority (1997) *The Financial Services Authority–An Outline.* London: FSA, p 33).

11 See Levi, M. (1996) Money Laundering: Risks and Counter-measures. In Graycar, A. and Grabosky, P. (eds) *Money Laundering in the 21st Century: Risks and Counter-measures.* Canberra: Australian Institute of Criminology.

12 Richards, J.R. (1999) *Transnational Criminal Organizations, Cybercrime, and Money Laundering: A Handbook for Law Enforcement Officers, Auditors, and Financial Investigators.* Boca Raton, FL: CRC Press, p 180.

13 Ibid, p 184.

14 Pinner, G. (1996) Money Laundering: The State of Play. In Graycar and Grabosky, op cit, p 16.

15 Non-committed or 'don't know' responses are not used in text or tables unless otherwise indicated.

16 Levi, op cit, p 11.

17 Gill and Taylor (forthcoming b), op cit.

18 Of the 143 respondents who disclosed between one and ten suspicious transaction reports (STRs) to NCIS, 58% (N = 83) did not use IT to monitor transactions while 42% (60) did; conversely, of the 94 respondents who submitted 11 or more STRs in an average year, 36.2% (34) did not use IT while 63.8% (60) did.

19 See also Gold, M. and Levi, M. (1994) *Money-Laundering in the UK: An Appraisal of Suspicion-Based Reporting*. London: Police Foundation.

Chapter 7

Trafficking in Women for the Purpose of Sexual Exploitation: Knowledge-based Preventative Strategies from Italy

Andrea Di Nicola[1]

This chapter proposes a method to prevent trafficking in women for sexual exploitation. It emerged from a study of the opportunities, resources and technologies which traffickers/exploiters use to conduct their illegal enterprises, the research being carried out on judicial cases in Italy. An important result of the analysis, also in terms of prevention, is the recognition of the existence of 'legitimate' actors who facilitate the trafficking activities, eg taxi drivers, hotel owners, and travel and employment agencies. Their actions, which often result in crimes at work, are among those that must be monitored and prevented.

Introduction

This chapter is an attempt to apply a 'knowledge-based crime prevention approach' to trafficking in human beings for the purpose of sexual exploitation to Italy. In attempting this, the following section defines the theoretical framework underpinning the approach; the third section outlines the *modus operandi* of organised criminals, and describes the opportunities, resources and technologies they exploit to traffic Eastern European women to Italy; and the fourth section suggests, from the evidence generated, preventative measures.

Many studies have shed light on the trafficking of human beings into the European Union. One set of contributions concentrates on the modalities of this criminal activity and describes how the traffic is distributed and controlled by separate organised crime groups working together.[2] Some of these writers also argue that the application of business concepts could be useful in the analysis of the trafficking.[3] Another set of contributions, made up of surveys, concentrate their attention on the characteristics of the victims.[4] Finally, a small volume of research focuses on quantifying the problem and estimates the dimension of the trade.[5] All this work, notwithstanding its diversity, helps to highlight the dynamics of trafficking and suggests ways to reduce it. What it seems to lack is a methodology able to provide guidance in drafting prevention policies. The interventions presented are dictated by coherent logic, but do not offer a method that could help to boost their effectiveness.

This chapter proposes such a method, based on empirical research conducted at the 'Transcrime' Research Centre on Transnational Crime at the University of Trento between 2001 and 2002, and forming part of the project entitled *A pilot study on three European Union immigration points for monitoring the international trafficking of human beings for the purpose of sexual exploitation across the European Union* ('MON-EU-TRAF'), under a grant awarded by the European Commission as part of the '2000 Stop Programme'.[6]

Finally, a comment for those readers who may consider it to be unusual to find a chapter on this subject in such a book. The purpose will become clearer as the methods used by criminals emerge. It will be shown that for trafficking to be successful a range of workers and workplaces are involved, including for instance taxi drivers and hotel managers, while administrators help traffickers to obtain visas for women, and owners of clubs/exploiters deprive women of their earnings. Trafficking thus certainly qualifies as a crime at work, and at least one of the problems in some prevention efforts to date has been the lack of recognition of this point.

Preventing organised crime: a theoretical approach

Organising themselves to commit crime is a way in which offenders can maximise expected gains and minimise the risks of punishment,[7] in a process very similar to the organisation of any legal entrepreneurial activity.[8] The central elements of every illegal (or legal) organisation can be considered as: the exploitation of opportunities[9] offered in a given environment; the acquisition of resources; and the technology employed.

The term 'opportunity' refers to any permanent factor—whether legal (eg legislation that involuntarily creates incentives for crimes) or illegal (eg a constant demand for a type of illegal service)—which may facilitate the rise and development of a particular criminal activity, by increasing profits and/or minimising the risk of punishment.

'Resources' are any available psychological, economic or material asset with which criminals have to be equipped to perform their illegal activities:

> Some resources are part of the offender ... Together these could be regarded as the offender's 'core competencies' for crime. Other resources relate to facilitators ... and to the scope for collaboration with others.[10]

'Technology', a concept borrowed from economics, engineering and operational research,[11] is any instrument an illegal organisation uses to achieve its final and preplanned criminal goal. It can be regarded as the way in which the organisation carries out its activity or part of its activity, and refers to the methods and tools employed to manage the organisation and to reach its goals. A good example of technology is the method chosen by an organisation to acquire resources and to combine them efficiently to achieve its objective.

In order to disrupt a criminal organisation, three strategies are recommended: blocking or reducing opportunities for the criminal organisation; blocking or reducing the organisation's access to resources; and making the establishment of technologies more difficult for the organisation. It is these three approaches that are the focus of this chapter.

To highlight the opportunities, resources[12] and technologies used in each phase of criminal activity, it is helpful to consider 'how crime works', that is to understand the criminal's *modus operandi*. This is a prerequisite for developing effective policy responses.[13] There are strong links here with the situational crime prevention approach, which also relies on the collection of data and the analysis of criminal phenomena (scanning loopholes, weaknesses of criminal activities, opportunities, etc) in order to elaborate the most effective solution, and with similarly problem-oriented policing.[14]

Combating criminals by understanding the organisation: the Italian case

This section discusses the *modus operandi* of the traffickers of Eastern European women for the purpose of sexual exploitation in Italy, and then comments on the main opportunities, resources and technologies used by these criminals. The data is drawn from the results of the MON-EU-TRAF project, referring to the analysis of some 18 cases prosecuted by Italian judicial authorities between 1999 and 2001.[15] The sources are court proceedings and/or preliminary investigation documents, and interviews with public prosecutors and law enforcement officials who dealt with the cases.

In order to standardise the analysis, each case was studied using a 'model spreadsheet', composed of a series of questions seeking relevant quantitative and qualitative information on the trafficking process. The cases under scrutiny are not a 'statistical sample', but given the care taken in selecting them and the help received from law enforcement officials and public prosecutors in identifying the most recent and 'representative' examples, the results presented can be considered to provide a good picture of the Italian situation.

The modus operandi *of traffickers*

It is customary to divide the entire trafficking process into three stages: recruitment; transfer and entrance; and exploitation.[16]

The recruitment phase was similar in all the examples studied. The starting point is a country in Eastern Europe: Romania, Bulgaria, Albania, Moldova, Ukraine, Russia or the Baltic States. The first stage of the process is managed by the local people responsible for recruitment. They look for two kinds of victim: women who are already working in bars or public places as prostitutes but who want to go abroad to increase their earnings and improve their living conditions; and younger girls (who may even be minors) living in poor socio-economic conditions and without jobs, looking for legal employment elsewhere. Recruitment is often conducted by means of advertisements in local newspapers, and on occasion travel agencies are involved. The advertisements usually set specific requirements regarding appearance, social status and age; some ask for photographs. The types of jobs offered to these girls include au-pair, dancer, model, housemaid, waitress and air-hostess.

The above-mentioned two groups of women are treated in different ways. In the former case, the women are perfectly aware of the work awaiting them. What they do not know is that they will be working on the streets under the control of an 'owner', who will keep them in servitude and take all their earnings. In the latter, the women are persuaded to leave their countries with false promises of legal employment. In this case, the victims sometimes know the recruiter and trusts him/her because of their friendship. Thus women leave the country voluntarily and in possession of their own legal documents; these are then taken away by the criminals as soon as they cross the border. Recruitment may also involve kidnapping. In general, this happens when a girl does not agree to the proposal made by the recruiter. In some cases, particularly those involving Albanian criminals,[17] there is also evidence of another method: they become engaged to the girls before leaving Albania in order to reassure the victims' families.

On occasion the recruitment phase starts as a specific commission. In this case, there is a very close link between the owner of a sex club and the traffickers' criminal organisation. The women are treated as resources, with the traffickers being asked by the owner to supply a certain number of women to be employed in a club or clubs.

Moving on to the transfer and entrance phase, there are two main east-west routes exploited by traffickers to enter Italy. The first is used by traffickers from Ukraine (the recruiters), Slovenia, Yugoslavia and Italy. Ukraine, Russia, Moldova and the Baltic States are the preferred places of origin of the women being trafficked. The border used for illegal entry into the country is that between Italy and Slovenia, along the boundaries of the provinces of Trieste and Gorizia. This is a geographical 'loophole': forest paths are well hidden and difficult to access, and police controls have only recently been reinforced. The women are taken from these eastern countries to Budapest, which is an important gathering point. According to the results of investigations, the traffickers use various means of transport to Budapest (planes, trains, lorries) depending on the place of origin. After the women have passed through Hungary, they enter Slovenia (Maribor or Ljubljana), where they are lodged in houses (sometimes still under construction) while they wait to be smuggled into Italy. The traffickers use lorries when there are more than twenty women, while they prefer cars (or taxicabs) for small groups. The trafficked woman are driven at night to forested areas close to the border, where a *passeur* is waiting to take them across the border on foot through the woods. Once in Italy, the *passeur* hides the women in the forest, where they wait for other persons with cars to collect them. The journey continues along the Trieste–Venice highway. Later, at a prearranged service station, the women are handed over to another member of the criminal chain, who takes them to the Venice-Mestre railway station or other destinations (Bologna, Rimini, Rome and other cities in the centre/north of the country). Traffickers from Bulgaria are also reported to use this route. This evidence has been drawn from cases in the Trieste Public Prosecutor's office, and reveals a very well organised network of criminals operating in synergy to manage the entire trafficking process. The criminal organisation dealing with sexual exploitation effectively 'orders' a certain number of women, and once the victims have been found, the trafficking ring does the rest.

Numerous persons, each with a specific role, are involved. The organisation has a vertical structure, with a Slovenian group leading the criminal operations and a number of Slovenians, Yugoslavs and Italians working for it. In some cases, the use of fake documents has been reported, while in others victims have been legally introduced into the countries with a tourist visa, or a legal work permit obtained with the help of co-operating Italians and also complaisant entrepreneurs. In one case at Salerno, for example, an Italian farm owner was convicted of being a member of a criminal association but not as a trafficker or as an exploiter. He had provided fictitious work contracts to traffickers and their victims, which allowed them to be issued with legal work permits and to live in and move freely around Italy. The farm owner was paid between €1550 and €2330 for each contract.

Albanian crime groups manage the second east-west route, or more accurately two routes. Either they recruit women from among their own nationals in Albania, or they take women from other Eastern countries (Moldova, Romania, Bulgaria, Ukraine and Russia). The trafficking route passes through Moldova, Bulgaria, Romania (recruitment places), Hungary, Yugoslavia (Belgrade, Podgorica),

Montenegro (intermediate destinations), and thence to Albania—the collecting country, where the whole process is organised. Not all of these places, of course, are traversed in a single journey; the countries and cities involved depend on the origin of the victims, the schedule drawn up for the journey and the destination of the women recruited. No precise data are available on the means of transport used by traffickers: planes, cars, trains, boats and motorboats are reported to be used in different combinations. The last part of the journey, from Albania to Italy, is common to both trafficking routes. The main departure points in Albania are Valona and Durazzo, where well-established motorboat services provided by Albanian *passeurs* are used. The women disembark on the Apulian coast near the cities of Lecce, Brindisi and Bari. The involvement of the Albanian police has been alleged in several cases. Corruption is quite common as a 'related crime'. Court records show that other offences committed together with trafficking in human beings are drug trafficking, trafficking in arms and explosives, and money laundering. With reference to drug trafficking, its co-existence with trafficking in human beings is rare; criminals concerned in the latter tend to be specialised. As far as money laundering is concerned, notwithstanding the huge profits generated by people trafficking,[18] there have only been a few cases of money-laundering investigations connected to the traffic in human beings. This might be due—as public prosecutors stress—to the difficulty of making financial investigations in the field. Money transfers rarely involve banks; often the cash is transported by hand to the traffickers' countries of origin, or money remittance services are used.

The last phase is that of exploitation. The cases analysed show that exploitation in the Italian sexual market, unlike other European countries, is mainly conducted on the streets. Although the street is not the only place where the exploitation occurs, it seems to be a specific feature of the Italian market.

Once they have arrived in Italy, the women are transferred to their destination. The means of transport depends on the distance involved. In general, exploiters prefer to use trains, but sometimes cars are used. The latter circumstance always involves the complicity of Italian citizens who are paid to drive the women using their own cars. In many cases, Italian taxi-drivers are involved. For instance, in a case from the Lecce Public Prosecutor's office, taxi-drivers were paid to transport women from the south to the north of Italy.

At the destination, the victims are lodged in flats rented by the criminal organisation or owned by its network, or in hotels. The managers of these hotels are very often perfectly aware of the criminal activities being conducted on their premises. For instance, they sometimes rent out rooms in the full knowledge that the documents they see are counterfeit, or while authentic do not match the person who has been registered. In some cases these persons have been convicted; in others, although the public prosecutors were reasonably convinced of their connivance, there was not enough solid evidence to take the case to court.

Apparently the victims do not pay directly for their board and lodging. Albanian women are usually forced to hand all their earnings over to their exploiters. All costs of board and lodging are thus met by the exploiters out of the earnings from prostitution.

Living conditions are very bad. The women are forced to work continuously for many hours. This depends on the exploiter, but is usually ten to 12 hours per day for 30 days a month. Working time is sometimes fixed: from nine or ten am to ten pm, or from ten pm to five am. The worst situation reported was one where victims had to work from nine am to one am. Illness is not considered an excuse (or permitted). If the women refuse to work, for whatever reason, they are beaten, and they receive the same treatment when they do not earn enough or do not want to give all their earnings to those who exploit/control them.

The women are forced to prostitute themselves on the street. The exploiter assigns each woman a specific place in a chosen area, which becomes the victim's workplace. The women may travel to these places on their own by train, or they may be driven there in the charge of a member of the criminal group. The former circumstance applies to victims with a certain degree of freedom, those that are trusted or are under extreme threat from the exploiter(s); the latter concerns situations in which closer control is exerted over the victims. Once at their workplaces, the women wait for clients, with whom they have sexual relations. The pimps usually establish the price of each kind of sexual service in advance, and this varies with the duration of the service. In the case of pregnancy (two cases out of 18, both involving Albanians), the women were persuaded, or rather forced, to abort. In some cases, abortions were performed illegally by members of the group, and caused serious injuries to the victims. The criminals were also convicted of abetting and/or practising illegal abortions.

Well-structured crime groups make no distinction between members in respect of who is in charge or in control. Flexibility is very high. In the Italian-foreigner organisations, especially Italian-Albanian, control is almost always delegated to the Italian component. In one case, the main criminal organisation paid for control services from the Italian section of the network (which was not fully part of the crime organisation); this section provided close control over the prostitutes, the area in which they worked, the allocation of a street pitch to each prostitute, and other connected services, such as the distribution of condoms.

The seizure of the victim's identity documents is a first means of control. After they have been deprived of their documents, the victims are provided with false documents and new identities, which are changed from time to time in order to avoid recognition by the police during checks. The use of violence and menaces is another instrument of control, because it undermines the victim's self-confidence. These are general methods adopted in almost all cases, but exploiters also resort to other devices, such as driving the victims to their workplaces, protecting them while working, fixing the prices of services and giving the women a certain number of condoms. The role of the controllers is very important in these cases because they are in constant contact with the victims by mobile phone, to warn them of problems and to collect all the daily earnings.

Opportunity makes ... the traffickers

In the different phases of the trafficking chain, organised criminals take advantage of existing legal and illegal opportunities.[19] This means that their decision to become traffickers and the way in which they organise their illegal activity are obviously influenced by external facilitating factors.

With reference to the recruitment phase, the women's desire to move to richer countries where they believe they can find better lives and economic conditions represents one of the greatest opportunities for criminals. According to the testimonies of the women in the proceedings analysed, they leave to improve their standard of living or that of their families. The welfare differential between EU member states and the origin countries of these women is the main push factor in their decision to abandon their countries, and this is exploited by criminals.

Law and regulation can also create incentives for crime. For instance, again in the recruitment phase, a further opportunity that seems to foster criminal activities is represented by the small number of options for legal migration in receiving countries. With high unemployment rates across the continent, Western European countries have enacted legislative restrictions on immigration. When it is (or women perceive it to be) very difficult to migrate legally, it is easy for organised criminals to step in and deceive them.

When we progress to the transportation and entrance phase, criminals seem to profit from the following opportunities: the ease with which certain types of documents can be counterfeited (judicial evidence shows that passports and identity documents from various Eastern European countries are very often false); the possibility, for citizens of a number of countries, to enter Italy or other transit countries as tourists, often without visas (after that, the women simply over-stay); the difficulty of patrolling some borders, due to their geo-morphological nature; the presence of 'hot gathering spots' and 'hot transit spots' (there exist countries where traffickers concentrate their victims or which are preferred as transit points—the criminals' reasons may differ, but they certainly exploit such points as offering better platforms at lower risk); and the high corruptibility of public servants, law enforcement officials and embassy/consulate staff in certain countries.

The exploitation phase is facilitated by two main factors. First, the high and constant demand for sexual services. This demand assures a permanent profit for criminals and feeds the supply. The demand comes from clients themselves, and from the owners of bars or sex clubs as intermediaries. As already mentioned, the entire trafficking process often starts as the result of a commission, a request for pretty, young girls.

The second factor is the connivance and help received from taxi drivers, owners and/or managers of hotels, landlords who rent houses and apartments, and the owners of legal commercial establishments. These 'legal facilitators' work for profit and turn a blind eye to the criminal activities, which they thus aid and abet. On occasion some are convicted, but it also happens that when charged a lack of evidence leads to their acquittal. The phenomenon of these legal facilitators must be carefully studied; dangerous forms of crime at work are involved, and are essential to traffickers/exploiters in minimising their risks and developing their criminal activities.[20] There is no doubt that if these services were not provided to the traffickers/exploiters, the whole organisation of the trade would become much more complicated.

Equipping the enterprise: the resources of the traffickers

Nothing can be done without women. In every enterprise, raw materials are essential. It seems very cynical to say so, but in this case the main raw materials are the girls. Girls are therefore the principal resource and the less available they are, the more difficult the entire criminal enterprise becomes.

In the transfer and entrance phase, there are other assets crucial to the performance of the illegal activity (and some of them do not belong directly to the core organisation, which has to co-operate with other groups or individuals to obtain them), namely: knowledge of transit routes; the capability of using threats, physical violence and rape; skills in the falsification of documents; skills in the corruption of law enforcement officials; availability of trucks or motorboats; 'aiding and abetting' resources (*passeurs*, taxi drivers, and other individuals who give food and lodging to the trafficked women at different places along the route provide traffickers with the essential resources they need to manage the transportation phase).

Some of the resources mentioned above are also required in the exploitation phase. In fact, the most important assets needed in this phase are: a work-force to control the women; availability of means of transport; availability of counterfeited documents; capability of using threats, physical violence and rape; lodging, and other 'aiding and abetting' resources.

Managing the enterprise: the technologies of the traffickers

At least four different types of technologies can be recognised in the *modus operandi* of the traffickers: technologies used to recruit women; outsourcing; vertical interdependencies; and control technologies.

As seen earlier, the main modalities through which contacts with girls are established during the recruitment phase are advertisements in local newspapers and the use of travel and employment agencies. The more developed these technologies, the more the supply of women will be constant and wide-ranging. The complexity of such work-methods is an indicator of the specialisation of the criminal group. Again, these behaviours may be seen as belonging to the area of crime at work.

The outsourcing of single criminal tasks to other criminals or criminal organisations is an essential method of managing trafficking activities. It occurs when human and/or material resources are drawn from outside the criminal group. It is common for the criminal enterprise to retain the core business, but to allocate some parts of the activity to other criminals who are more skilled or more prone to risk-taking. This outsourcing happens especially during the transfer and entrance phase, but it is not uncommon for it to occur during the other phases.

In the transfer and entrance phase, vertical interdependence between crimes must also be mentioned. It occurs when organised criminals or other criminals, with the aim of committing an offence, also commit a series of intermediate or instrumental crimes. In order to finalise a crime of particular importance (in terms of effects or gains), a chain of offences are committed. In the case of trafficking in human beings for sexual exploitation, a series of crimes are committed and are linked together in order to gain criminal profit from the exploitation. These include, among others: the counterfeiting of documents; the corruption of public officials; the use of complaisant citizens who aid and abet traffickers by offering food and lodgings to the women; and the violation of laws concerning legal migrations.[21]

Among the most important technologies used in this phase are those used to control the women. Worthy of note is the practice of supplying them with a fixed number of condoms in order to ascertain the number of times sexual services have been provided. The method of fixed prices and of providing a certain number of condoms enables the exploiters/controllers to detect any theft of the money earned during the day by the prostitutes.

Interventions to tackle the traffickers

As a consequence of the above analysis, it is possible to suggest preventative interventions aimed at: reducing/blocking opportunities; making it more difficult or riskier to obtain useful resources to carry out each single phase of the criminal activity; making it more difficult or riskier to establish the technologies necessary to conduct each single phase of the criminal activity.

It is of course seldom easy to find solutions to a criminal problem, especially if they are not the usual criminal law responses. The strategies proposed here are not exhaustive, but they will hopefully raise awareness of the possibilities. They are tailored to the peculiarities of what occurs in Italy, and it is likely that they may not be readily applicable to other situations.

Techniques for reducing/blocking opportunity could include the following:

For the recruitment phase:

- finding economic alternatives for women, and strengthening job opportunities:

> Numerous intergovernmental organisations such as UNICEF and NGOs in the countries of origin are actively involved in providing such opportunities to children and young women to prevent their departure ...[22]

- when enacting or choosing policies that have an effect on immigration, taking into account that they might create incentives for criminal behaviour and considering measures to reduce these incentives; with respect to this, the placement of crime-proofing mechanisms in the legislative process could be examined.[23]

For the transfer and entrance phase:

- increasing the difficulty for the criminal of falsifying identity documents (making them more foolproof by, for instance, changing the format), and equipping law enforcement officials with improved, up-to-date and standardised means of identifying false documents;[24]

- designing an effective visa policy regime, which clearly takes into consideration 'risk categories of applicants';[25] different and stricter visa issuance regimes could be applied to countries involved in the trade:

> Visa granting should be followed by control in destination countries. Measures such as a list of indicators could be implemented in order to prevent traffickers from using this system.[26]

- more effective use of technical devices to exercise better control over weak borders;[27]

- acting on specific 'hot gathering spots' and 'hot transit spots' (countries or cities), through policies aimed at developing social and economic progress; some of the countries concerned are likely to enter the European Union in 2003 or 2004, and could benefit from EU Structural Funds to increase their level of legality and security, with these EU-financed strategies being tailored to their specific needs;[28]

- the development of anti-corruption strategies, especially amongst embassy and border guard officials; this could also mean reducing their discretionary powers and increasing their responsibility through ad-hoc reforms.[29]

For the exploitation phase:

- intervening in the demand for prostitution in order to reduce it; this would mean looking for methods capable of discouraging the use of prostitutes' services; these methods could be varied but 'rough and ready' studies should be make better profiles of the demand and consequently act. On this point, a Home Office research report recommends 'increasing the awareness of trafficking in women and the gross violations of human rights involved', a consideration of 'the possibility of targeting customers' and increasing 'research and knowledge about men as customers/buyers';[30]

- devising strategies (also based on a classic situational prevention approach or on environmental design) to make street prostitution less easy to practise.

Among techniques for reducing/blocking access to resources, one could suggest the following.

For the recruitment phase:

- making women less vulnerable through the design of serious and effective information campaigns (in terms of numbers of women contacted) to make potential victims aware of the risks of trafficking and of the legal possibilities of migration. These campaigns should be implemented in those countries where the victims of trafficking and exploitation originate. More in-depth studies should be carried out to determine the ages and characteristics of victims, so that the campaigns can be properly targeted. The means of transmitting these campaigns should be selected according to the features of potential victims. Information campaigns are already carried out by various NGOs and other international organisations in different parts of the world—what is needed is co-ordination, and solutions tailored to the type of victims and to their needs.[31]

For the transfer and entrance phase:

- designing new forms of controls over boat owners and taxi drivers (in this latter case something more than the licensing that already exists), and considering possible and speedier means of seizing and confiscating boats and taxis, without the necessity of a criminal trial; this is a first area of 'crime at work' where interventions should be studied;

For the exploitation phase:

- improving the administrative control systems for hotel management, and in general more 'licensing and monitoring [of] the premises that are likely to be used'[32] by traffickers; this is a second 'crime at work' area where measures are highly recommended.

Among the techniques aimed at making it more difficult and risky to establish technologies, one could suggest the following.

For the recruitment phase:

- devising methods to neutralise the effects of newspaper advertisements searching for women in origin countries; a simple way could be to buy advertisement space, in the same newspapers and on the same pages, warning women about the risks of being trafficked;

- stricter controls on travel and employment agencies (a third relevant sector of 'crime at work'); one method could be to make them subject to an administrative/governmental authorisation to operate, to obtain which the agencies would have to pass careful administrative checks; citizens should be informed of this authorisation and encouraged to make use only of authorised agencies.

For the transfer and entrance phase:

- identifying areas in which criminal outsourcing takes place, and disrupting them; this is a sector where criminal law is more applicable, but this does not mean that preventative solutions cannot be sought and implemented.

For the exploitation phase:

- distributing free condoms among prostitutes on the street (for instance, by the social services or NGO personnel); this could be a simple way of disorganising the working methods exploiters use to control their victims; as mentioned above, exploiters supply the girls with a fixed number of condoms, and decide the price of each sexual service; by simply counting the number of condoms remaining, they are able to calculate what the woman should have earned.

The study of opportunities, resources and technologies shows that preventative efforts should be concentrated on particular areas: on the *victims* (in origin, transit and destination countries); on *legislation* and on *visas*; on *identity documents*; on *borders*; on *weaknesses in origin and transit countries*; on *demand for prostitution*;

on *legal facilitators (entrepreneurs)*; and on *co-operation among different criminal groups*. These key areas require study and imaginative preventative measures. This article has attempted to indicate a methodology that could be used to design and orient these measures.

Notes

1 Andrea Di Nicola is a researcher at Transcrime, Research Centre on Transnational Crime, University of Trento; email: andrea.dinicola@transcrime.unitn.it. He is grateful to Silvia Decarli, a researcher at Transcrime, with whom the final and Italian reports of the research were drafted.

2 See, for instance, Limanowska, B. (2002) *Trafficking in Human Beings in Southeastern Europe*. Belgrade: UNICEF; Europol (2000) *Trafficking in Human Beings: General Situation Report 1999*. The Hague: Europol; International Organization for Migration (2000) *Migrant Trafficking in Europe: A Review of the Evidence with Case Studies from Hungary, Poland and Ukraine*. Geneva: IOM; Salt, J. (2000) Trafficking and Human Smuggling: A European Perspective. *International Migration*. Vol. 38, No. 3, pp 41–45.

3 Salt, op cit, p 49; Salt, J. and Stein, J. (1997) Migration as a Business: The Case of Trafficking. *International Migration*. Vol. 35, No. 4, pp 467–94; Aronowitz, A.A. (2001) Smuggling and Trafficking in Human Beings: The Phenomenon, the Markets that Drive it and the Organisations that Promote it. *European Journal of Criminal Policy and Research*. Vol. 9, No. 2, pp 163–95.

4 See, for instance, International Organization for Migration and International Catholic Migration Commission (2002) *Research on Third Country National Trafficking Victims in Albania*. Tirana: IOM; International Organization for Migration (2001) *Victims of Trafficking in the Balkans: A Study of Trafficking in Women and Children for Sexual Exploitation to, through and from the Balkan Region*. Geneva: IOM, pp 1–8.

5 International Organization for Migration (1996) *Trafficking in Women to Austria for Sexual Exploitation*. Budapest: Migration Information Programme; Kelly, L. and Regan, L. (2000) *Stopping Traffic: Exploring the Extent of, and Responses to, Trafficking in Women for Sexual Exploitation in the UK*. Police Research Series, No. 125. London: Home Office; Carchedi, F., Picciolini, A., Mottura, G. and Campani, G. (eds) (2000) *I colori della notte: migrazioni, sfruttamento sessuale, esperienze di intervento sociale* ('The colours of the night: migrations, sexual exploitation and experiences of social intervention'). Milan: Franco Angeli.

6 The project was conducted (from south to north) in Spain, Italy and Finland; its objective was to provide knowledge useful for the devising of effective policies to prevent trafficking in human beings for the purpose of sexual exploitation in Europe. Partners in the research were the National Research Institute of Legal Policy (Finland) and the Research Centre on Criminology in the University of Castilla-La Mancha (Spain). The final report is available on the Transcrime website, at *http://www.transcrime.unitn.it*.

7 McIntosh, M. (1975) *The Organisation of Crime*. London: Macmillan, p 14.

8 Bernard, C.I. (1938) *The Functions of the Executive*. Cambridge, MA: Harvard University Press; Williamson, O.E. (ed.) (1990) *Organization Theory: From Chester Bernard to the Present and Beyond*. Oxford: Oxford University Press; Hatch, M.J. (1997) *Organization Theory and Theorizing: Modern, Symbolic-Interpretive and Postmodern Perspectives*. Oxford: Oxford University Press.

9 On this issue, see Clarke, R.V. (1997) *Situational Crime Prevention: Successful Case Studies*. 2nd edn. New York: Harrow and Heston, pp 12–26.

10 Ekblom, P. and Tilley, N. (2000) Going Equipped: Criminology, Situational Crime Prevention and the Resourceful Offender. *British Journal of Criminology*. Vol. 40, No. 3, p 382.

11 Hatch, op cit, pp 127–60.

12 'To highlight resources in the causation of crime is, by the same token, to suggest important generic means of prevention. ... Awareness of resources can ... suggest means of restricting their availability and use ... a kind of offender incapacitation' (Ekblom and Tilley, op cit, p 387).

13 Cornish, D.B. and Clarke, R.V. (2002) Analysing Organised Crime. In Piquero, A.R. and Tibbetts, S.G (eds) *Rational Choice and Criminal Behaviour*. New York: Routledge, pp 41–63.

14 Clarke, op cit.

15 All the cases are taken from criminal proceedings that were at different stages in the judicial process. There were cases under preliminary investigation, cases committed for trial, and cases already adjudicated. Seventeen come from Italian Public Prosecutor's Offices (one from Brescia, two from Busto Arsizio, ten from Lecce, one from Salerno, one from Trieste, two from Turin) and one from the Department of Public Security of the Italian Ministry of the Interior.

16 Salt and Stein, op cit; Europol, op cit.

17 For an in-depth analysis of Albanian criminality, see Direzione Investigativa Antimafia (2001) *Criminalità albanese in Italia* ('Albanian criminality in Italy'). Rome: mimeo.

18 According to the results of the MON-EU-TRAF project in Italy, the estimated total number of victims trafficked for sexual exploitation in the period between 6th March 1998 and 31st December 2000 ranges from a minimum of 7260 to a maximum of 14,520. The yearly estimated average of victims (for 1999 and 2000) ranges from a minimum of 2640 to a maximum of 5280. The yearly estimated turnover from the sexual exploitation of trafficked people for both 1999 and 2000 ranges from a minimum of €380.16–760.32 million to a maximum of €475.20–950.40 million.

19 With reference to criminal opportunities, see Felson, M. and Clarke, R. (1998) *Opportunity Makes the Thief: Practical Theory for Crime Prevention*. London: Home Office.

20 The connivance of legitimate actors who have a 'control role' is not only an opportunity that reduces risks, but also a resource. It will therefore also be put in both categories.

21 Adamoli, S., Di Nicola, A., Savona, E.U. and Zoffi, P. (1998) *Organised Crime Around the World*. Publication Series No. 31. Helsinki: HEUNI ('European Institute for Crime Prevention and Control'), pp 16–18.

22 Aronowitz, op cit, p 186.

23 This would mean inserting crime risk assessment mechanisms into the legal decision-making process. In Sweden and other Scandinavian countries, these mechanisms, though rudimentary, are already in place. The discussion on this issue is high on the agenda of many EU countries. See Max Planck Institute for Foreign and International Criminal Law (2001) *Criminal Preventive Risk Assessment in the Law-Making Procedure*. Final report on research project 1999/FAL/140, funded by the EU 'FALCONE' programme and the Max Planck Institute. Freiburg: Max Planck Institute.

24 On this issue, see Europol, op cit, p 46.

25 Ibid.

26 Ibid.

27 Effective control of the border is a central topic in the EU debate. See Guardia di Finanza (2002) *Feasibility Study for the Setting up of a European Border Police*. Rome: GdF.

28 The use of EU Structural Funds to promote legality and security has only been experimented with in the least developed regions of Italy, based on the idea that insecurity is a factor leading to low economic and social development. Promoting security and legality would instead start a virtuous circle of development. During the EU's 1994–1999 programming period, the operational programme 'Security and Development in the South of Italy' was financed by the EU Commission and, given the positive results, refinanced during the 2000–2006 period. This experience could be exportable.

29 On this issue, see the results in the report (unpublished) of the Regional Seminar on Judicial Networking against Corruption and Trafficking in Human Beings, organised by the Council of Europe and the Office of the Government of the Republic of Slovenia for the Prevention of Corruption, 19th–22nd June 2002, Portoroz, Slovenia.

30 Kelly and Regan, op cit, p 39.

31 For instance, IOM organises mass information campaigns in countries of origin. See, as an example for rural areas in the Balkans, IOM (2001) *IOM Counter Trafficking Strategy for the Balkans and the Neighbouring Countries*. Geneva: IOM, p 5.

32 Europol, op cit, p 46.

Chapter 8

Stealing from Shops: A Survey of the European Dimension

Joshua Bamfield[1]

A short self-report questionnaire was used to discover key shrinkage, theft and loss prevention data from major European retailers. Valid responses received totalled 424, from retailers in 16 countries with a combined turnover of 22 per cent of Europe's retail trade. Results were reported mainly by country. This survey showed that the average rate of shrinkage was 1.42 per cent across large European retailers, although national results varied considerably. The total cost of crime for the retail sector (stolen property plus retail security costs) was estimated to cost European retailers €29.599 billion.

Stores considered that customer thieves were the biggest problem they faced, although retailers showed increased concern about the cost of theft by employees. Larger retailers, particularly those operating large stores, spent the equivalent of 25 per cent of total crime-related losses on retail security.

Irrespective of country, there was considerable similarity in the main types of merchandise stolen. Retailers in every country also expressed considerable concern about the unwillingness of their criminal justice system to deal with shop thieves in a manner which deterred re-offending. They complained that a lack of interest by many police forces in arresting and processing customer thieves where small values were involved undermined retailers' loss prevention strategies.

Introduction

Although the last ten years have seen a number of research reports into the level of retail crime in Britain and America,[2] there have been comparatively few retail crime reports carried out in mainland Europe,[3] and practically no inter-country research.[4] The task of managing security, like every other area of management, must necessarily be based on accurate data to assess progress and to make comparisons with like organisations. It is conceivably of interest to retailers, academics and public authorities to know how much retail crime there is in Western Europe, and to be aware of the major differences between countries.

The research published in this chapter was carried out for the first of what was intended to be a series of reports published every six months about retail crime levels and crime issues in Western Europe, entitled the *European Retail Theft Barometer*. This chapter provides information on the first such report.[5] It was sponsored by Checkpoint Systems, a supplier of electronic article surveillance equipment. The research and the preparation of the report were carried out by the Centre for Retail Research (CRR), with no interference from the sponsors.

The geographical area chosen for the survey was Western Europe. The economies of the countries concerned were all relatively advanced, and retailers had operated in free markets since the early 1950s, although this was only true of part of Germany. The systems and procedures used by retailers in different countries, whilst not identical, were thought to be similar enough to enable valid conclusions to be drawn from the data provided. The sixteen countries were Austria, Belgium (including Luxembourg, for data analysis purposes), Denmark, Finland, France, Germany, Greece, Ireland, Italy, the Netherlands, Norway, Portugal, Spain, Sweden, Switzerland and the UK. The retailers approached for data were large companies, in the belief that such organisations, operating in competitive markets, would be more likely to have accounting and loss-prevention systems and procedures that were similar to one another's; it was also felt, based on the author's previous research into UK loss prevention activity,[6] that larger companies were more likely to have effective reporting mechanisms, enabling loss prevention data to be collected centrally.

The sixteen countries are increasingly an economic unit. Even though not all countries in Western Europe are members of the European Union or have adopted the Euro as their currency, there is tariff-free trade across Western Europe for goods and services and there are few restrictions on currency flows or travel, or on the entry of foreign companies into domestic markets. There are already a number of European international store chains, such as IKEA, Aldi, Tesco, Netto, Lidl, Albert Heijn, H&M and Carrefour, which trade in different countries, as well as US chains such as Toys'R Us and Walmart.

This study is one of the largest surveys of retail crime, and the first to attempt to measure differences in retail crime costs between several countries. Every country was assessed by the same survey instrument and the same methodology, and the research was carried out at the same time, thus overcoming a common problem in making inter-country comparisons based on different types of sample, different questionnaires and different methodologies. The quality of the data may vary considerably: nonetheless, this is the first exercise of its kind, on which others may build.

Method

This survey mainly concerns 'shrinkage' losses (which can be generally defined as loss of inventory caused by crime, wastage and error). It deals solely with theft by customers, staff and suppliers (plus administrative error affecting shrinkage) and not other types of crime against shops such as robbery, violence, arson, terrorism, virus attacks or major frauds. British data for 2000 shows that theft represented 93 per cent of all crime losses suffered by retailers.[7] Although this figure may not be valid for other European countries, it does at least indicate that the shrinkage approach may nevertheless provide data relating to the larger proportion of retailers' crime losses.

The survey tool adopted was a self-report questionnaire, which was sent to the head offices of those major retailers whose performance and attitudes were being sampled. Respondents were asked whether they were reporting results for a group or a single operating unit. Although it might have been possible to collect data from individual stores or outlets, the cost of covering sixteen countries in this way would have been prohibitive (approximately €400,000), and prior permission from head office would in any event have been required before the stores could have been approached.[8]

A great deal of attention was devoted to reducing the barriers to a recipient completing the questionnaire. The questionnaire was sent to a named individual in the head offices of 1550 large retailers. In order to encourage a high response rate, a short questionnaire was used, capable of being completed in 15 minutes and consisting of information that a competent security manager would already have available. (Shrinkage data, whatever its imperfections, is kept readily to hand by security managers.) To create interest in completing the questionnaire and to improve credibility, local publicity about the study was generated in the retail trade press of the different countries.

The questionnaire was available in five languages: English, French, German, Italian, and Spanish. Questionnaires were sent out in September 2001, with a covering letter to the named finance directors or security managers in 1550 of the largest Western European retailers. In bi-lingual countries, companies were sent a questionnaire in the language which was most closely associated with the individual's name or company title. Dutch and Nordic retailers were asked to complete forms in English. Non-respondents received two further letters. The sample of retailers was obtained from several sources, including the CRR's own database, international trade directories and external databases. The sample was a structured random sample. It was structured both horizontally and vertically. The number of retailers from each country was broadly proportional to the relative size of its retail sector (horizontal structure) measured by sales turnover, although small countries were over-sampled to ensure a more representative response (see

Appendix). The vertical structure meant that across Western Europe as a whole it was necessary to ensure that all the main retail trades or 'vertical markets' were being sampled—ie florists as well as food retailers, sports shops as well as department stores. Thus, in the Netherlands, for example, a sample of large retailers was chosen reflecting the main vertical markets.

The questionnaire could be returned by fax, post or email, or completed on-line. There were 424 useable questionnaires returned, and a further 45 were sent back because the named person had moved to another company or was unwilling to complete the survey. This was a response rate of 27.4 per cent. The 424 retailers had a combined turnover of €449.8 billlion,[9] 30,243 retail stores and about 22 per cent of the entire retail business of Western Europe. The Appendix provides further information about the respondents. Less than ten responses were received from Portugal (nine), Greece (eight) and Sweden (two), which means that the data for these countries may not be robust. For completeness, however, these countries' results have been included along with the rest of the sample.

In order to calculate the results, the data were weighted in accordance with the turnover of the company concerned. This method was used to prevent an unrepresentative result occurring from the use of a simple arithmetical average. The results for each country have been calculated with company figures grossed up, using weightings allocated in proportion to the size of the different vertical markets (eg grocery, clothing, department and general stores). At a European level, each country's figures have been weighted in relation to that country's aggregate retail turnover (measured in euros) as a proportion of total European retail turnover. This survey methodology assumes that shrinkage and loss prevention data of non-participants and smaller retailers are similar to those reported.

There are traditional difficulties in mounting retail crime surveys. The key one is that retailers cannot know how much crime they have suffered, but only know their shrinkage, which includes an element of non-crime losses. The rate of response may be poor, because retail security managers can be reluctant to give sensitive information to persons outside the company. Retailers, surprisingly, may honestly not know data about the numbers of persons arrested or how many security guards they employ. In addition, there can be differences in accounting systems as to the definition of shrinkage, and thus apparent differences in shrinkage may be the product of different methods of accountancy as well as of differences in actual losses.

Shrinkage levels in Western Europe

Retailers were asked to indicate their sales turnover and their levels of shrinkage as a proportion of turnover. Retailers only very rarely know with complete confidence the total value of crimes committed against them. The most recent British Retail

Consortium estimate is that only 22.2 per cent of retail theft losses are 'known', the remaining 77.8 per cent being calculated by reference to shrinkage.[10] This is because only a minority of crimes are witnessed at the time or detected later: the value of losses retailers face over a period is normally calculated from estimating shrinkage (or loss of inventory). This inventory loss also includes non-criminal losses from wastage, pricing errors, and a range of identification or accounting failures.

The average shrinkage rate as a percentage of turnover of the major retailers in the fifteen countries was 1.42 per cent in the period 2000–01. Table 1 shows that the incidence of shrinkage varied considerably. The UK (1.76 per cent), Greece (1.73 per cent) and France (1.59 per cent) had the highest rates, whilst Switzerland (0.87 per cent), Austria (0.98 per cent), and Sweden (1.19 per cent) had the lowest rates. (It is worth noting that the German-speaking countries, Switzerland, Austria and Germany itself (1.21 per cent) formed a group with a very low incidence of shrinkage.) The total value of all this shrinkage to Europe's retailers was €28.906 billion.

Table 1. Shrinkage (as % of turnover), 2001 and 2000

Country	2001	2000	% change
Austria	0.98	0.93	5.4%
Belgium/Luxembourg	1.28	1.35	-5.2%
Denmark	1.21	1.26	-4.0%
Finland	1.44	1.41	2.1%
France	1.59	1.53	3.9%
Germany	1.21	1.29	-6.2%
Greece	1.73	1.59	8.8%
Ireland	1.24	1.21	2.5%
Italy	1.30	1.24	4.8%
The Netherlands	1.42	1.38	2.9%
Norway	1.56	1.41	10.6%
Portugal	1.38	1.30	6.2%
Spain	1.53	1.45	5.5%
Sweden	1.19	1.28	-7.0%
Switzerland	0.87	0.91	-4.4%
United Kingdom	1.76	1.81	-2.8%
Totals	**1.42**	**1.40**	**1.40%**

The European costs of retail crime

The above €28.906 billion shrinkage loss includes a proportion caused by administrative or internal error, ie which does not stem from crime. To establish the retailer sector's crime losses, therefore, we need to deduct from the shrinkage figure an estimate of the losses caused by staff errors, mistakes and poor administration. Retailers estimated that 'internal error' as opposed to crime was an average of 17.6 per cent of total shrinkage. If this is correct, it means that 82.4 per cent of shrinkage, or €23.818 billion, was thought to be caused by crimes against their shops. The estimate reflects the loss of inventory caused by the theft of property and its equivalent in monetary terms, whether as direct theft, collusion, refund fraud or payments fraud. This admittedly crude approach is not ideal, but is really the only one possible because most retail crime is *unknown*, and the alternative, shrinkage, measures much more than direct and indirect crime.

To calculate the total cost to retailers of shop crime, we add the costs of retail security and loss prevention to the net losses (crime-related shrinkage) caused by crime (Table 2). This follows the procedure used by the UK British Retail Consortium's *Retail Crime Survey*. The total costs of security and loss prevention for 2000/01 were €5.781 billion or 0.29 per cent of retail turnover.

Table 2. The European costs of retail crime

	€ billions
Costs of total shrinkage	€28.906
minus internal error/administration	€5.088
equals net losses caused by crime	€23.818
plus costs of retail security	€5.781
***equals* total costs of retail crime**	**€29.599**

Thus the European costs of retail crime, made up of losses caused by crime and the costs of security, were €23.818 (total crime losses) plus €5.781 (security costs), a combined total of €29.599 billion. This sum represents a significant element in the cost of living: for example, it is equivalent to €76.83 for every man, woman and child in Western Europe. The cost structure of retail crime is unlike that of most domestic or business crime in that retailers not only suffer the costs of crime but also spend significant amounts on security and loss prevention—and as a result, the great majority of offenders who are apprehended are caught by retail staff, not the police. The additional €5.781 billion that retailers spend on security, in order to prevent crime and/or detect offenders, amounts to a further 24.3 per cent of their crime losses.

Sources of retail shrinkage

Retailers estimated that customer theft was the greatest security problem they faced, responsible for 45.7 per cent of all shrinkage (Table 3 and Figure 1);

employee or staff theft was thought to account for 28.5 per cent; and theft by suppliers or their staff 8.2 per cent. As noted previously, the proportion of shrinkage accounted for by internal error accounted for the remaining 17.6 per cent.

Table 3. Perceived sources of retail shrinkage in Europe, 2001

	Percentage of total shrinkage	Estimated cost in Western Europe
Customers	45.7%	€13.210
Employees	28.5%	€8.238
Suppliers	8.2%	€2.370
Internal error	17.6%	€5.088
Totals	**100.0%**	**€28.906**

These figures are estimates—respondents' perceptions—although they are the conclusions of many of the most experienced loss prevention managers in the industry. Because the figures deal with unknown as well as known theft, they may be inaccurate. However, our figure of 17.6 per cent for internal error is nearly identical to the estimate of 17.5 per cent given by the most recent report in the Hollinger series for the same category in US retailing,[11] although Beck and Bilby show a figure of 27 per cent for what they term 'process failure' in the fast-moving consumer goods sector (mainly grocery and branded dry goods and toiletries), and is likely to include several elements of wastage.[12] Horst gives a figure of 22.3 per cent for German retailing.[13]

Figure 1. Perceived sources of retail shrinkage in Europe, 2001

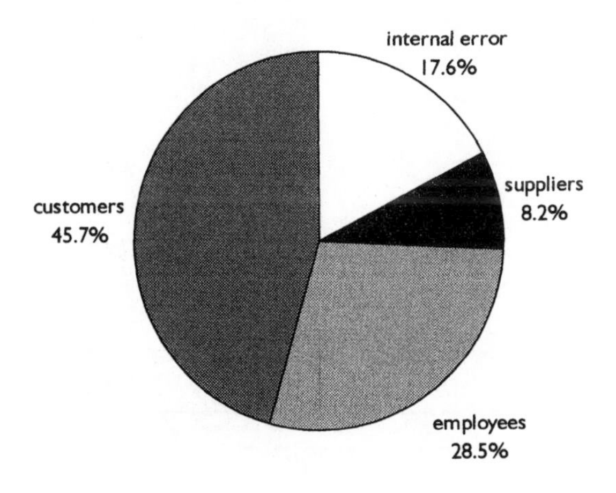

The estimate of 28.5 per cent of shrinkage caused by staff theft is more controversial. The most recent US study by Hollinger and Davis gives the percentage of retail shrinkage caused by staff theft as 45.9 per cent. Except for one year, the Hollinger series has always shown US staff theft exceeding the impact of customer theft.[14] In fact, the proportions of customer and staff theft for the survey and the European Retail Theft Barometer are almost the reverse of each other. Hollinger and Davis estimate customer and staff theft at 30.8 and 45.9 per cent respectively, while the European Barometer reports 45.7 per cent and 28.5 per cent respectively. Have European retailers somehow missed the fact that employees are a larger source of loss than customer thieves? Table 4 shows that retailers from every European country felt that customer theft had a greater impact than staff theft on losses; but the proportions attributed to staff theft varied greatly between countries. Retailers in the UK, Italy, Greece, Denmark and the Netherlands believed they suffered a high incidence of staff theft, whilst stores in Austria, Germany, Ireland, Portugal and Spain claimed to have the lowest proportion of shrinkage caused by such crime. Obviously, there may be cultural reasons why security managers in certain countries might feel unwilling to investigate or to admit to high levels of employee theft. But by the same token, culture or upbringing may also inhibit theft amongst the workforce.[15] Thus whilst there may be some under-reporting of staff theft amongst European retailers, it seems unlikely that current levels are comparable to those of the USA.

Table 4. Sources of loss for European retailers

Country	customers	staff	suppliers	internal error
Austria	51.3%	22.9%	7.5%	18.3%
Belgium/Luxembourg	44.1%	30.7%	7.4%	17.8%
Denmark	43.7%	32.0%	8.5%	15.8%
Finland	46.0%	30.6%	10.9%	12.5%
France	44.9%	27.1%	7.8%	20.2%
Germany	48.0%	22.5%	9.5%	20.0%
Greece	43.6%	31.4%	7.8%	17.2%
Ireland	51.6%	20.9%	13.2%	14.3%
Italy	40.6%	33.5%	7.7%	18.3%
The Netherlands	44.7%	31.2%	8.3%	15.8%
Norway	50.1%	24.9%	10.6%	14.4%
Portugal	54.5%	21.9%	6.3%	17.3%
Spain	49.5%	22.3%	10.2%	18.0%
Sweden	49.9%	25.3%	9.7%	15.1%
Switzerland	42.1%	35.2%	4.4%	18.3%
United Kingdom	41.6%	32.1%	2.3%	24.0%
Totals	**45.7%**	**28.5%**	**8.2%**	**17.6%**

Types of merchandise stolen

Retailers were asked to indicate which goods were most likely to be stolen, and the information collected covered all vertical markets. In this section, discussion will be restricted to two vertical markets, food/grocery (including hypermarkets and food superstores with a sales area greater than 4 000 square metres) and department stores. The responses of retailers were much the same, irrespective of the country on which they reported.

There were a few variations between countries, mainly those relating to strong local preferences: for example, Pernod or Ricard was a high-risk drink only in France. Tea was little stolen in the UK (perhaps because it is often a bulky low-price commodity line there) but highly desirable (costly and in small packages) in the Netherlands and Germany. Respondents provided a combination of brand names and product groups that were most likely to suffer theft, and these were common to retailers in most countries. They are listed in Table 5.

Table 5. Most-stolen items by type of retail business

Food and grocery	Department store
Food	Designer fashion
Alcohol	Designer accessories
Fresh meat	Small electrical items
Cosmetics/perfumes/fragrances	Womenswear
Tobacco	Sports shoes and sportswear
Shaving products and razor blades	Wallets
Perfumes/fragrances	Socks and scarves
Health and beauty	Gifts and toys
Baby products	Perfumes
Chocolate	Health and beauty
Video tapes/DVDs/CDs	Watches
Clothing	Jewellery
Tea and coffee	Small electrical goods
Electronic games	Computer software
Cellular/mobile telephones	CDs/videos/DVDs
Telephone cards	

Security managers felt that practically anything available in a shop was capable of being stolen, but goods with certain characteristics were the most vulnerable to theft. These characteristics were: high-value; relatively small-size; 'designer' brand, or manufactured by a well-known company; in great or regular demand by the public; and snob appeal, a cult, trend, or craze, particularly where supply was less than demand. Products meeting this specification could be slipped easily into the pocket or concealed about one's person, and readily sold direct to 'customers' or to an intermediary. Classic stolen products would be Sony Playstation games, nationally-marketed DVDs, Christmas 'must-have' toys, branded Scotch whisky, a highly advertised fine fragrance or toiletry, or the blades for the latest Gillette razor. These results provided little support for the idea that shop theft is mainly concerned with the poor stealing food to survive or with pensioners stealing tins of cat food for their pets.

Thieves apprehended and arrested

In 2001, European retailers arrested 1,266,555 criminals for theft or fraud, equivalent to the entire population of a substantial European city such as Milan or Birmingham. Most large food retailers reported that they apprehended between 6000 and 25,000 thieves every year and department and general stores, hardware and music stores also arrested substantial numbers. In contrast, many clothing chains or smaller non-food retailers, particularly those with only a handful of staff on duty or which operated from smaller stores, arrested only one or two shoplifters per store in 2001.

Of those apprehended in 2001, 1,201,571 were customer thieves, who stole an average of €51.20 per incident (Table 6). The number of employees apprehended was 64,984, who stole an average of €685.13 per incident. Apprehensions of staff were 5.1 per cent of total persons apprehended. It is significant that the average amount stolen by an apprehended member of staff was more than 13 times the average amount stolen by a customer thief.

The figures for customer thieves are likely to be an underestimate. Discussions about the results of the survey with 23 retailers in the UK, the Netherlands, Germany and France indicated that head offices may only be informed in 50–70 per cent of cases where people have been apprehended for shop theft; reports may not be submitted for young persons or for people who only steal small amounts, and a proportion of other cases may not be written up correctly or may fail to be sent to the head office because of work pressure, holidays or changing staff rotas.

The practice of 'banning' (or formally excluding) thieves from their shops was used comprehensively by UK retailers, many of whom banned more than 90 per cent of all thieves from their stores. Where retailers in other countries used banning, they usually did this only for the worst offenders; alternatively, they might offer the thief the option of either a store ban or being reported to the police. The practice of banning shoppers from stores was considered to be legal in only *some* of the countries surveyed, although no retailer reported that a ban had been successfully overturned by a court, implying that the limitations on banning might be self-imposed rather than judicial or constitutional.

Table 6. Apprehensions of thieves by European retailers

	Customers	Employees
Number arrested	1,201,571	64,984
Average value per theft	€51.20	€685.13
Outcomes		
Warned	51.9%	8.0%
Banned	23.8%	15.0%
Sacked (staff only)	-	92.0%
Referred to police	24.3%	53.0%

Although retailers reported different rates of shrinkage, crime and apprehensions, they were united across all the countries by concern about the criminal justice system in their own country. Some 64 per cent of retailers believed the sentences of the courts provided little deterrent to shop thieves. In dealing with thieves and deciding whether to involve the police, retailers claimed that they concentrated on the worst offenders, and that they wished to save their staff time as well as avoid harmful publicity for the store. Thus only 24.3 per cent of customer thieves and 53.0 per cent of employee thieves were reported to the police. Of those staff who were apprehended for fraud, 92 per cent were dismissed, in order to remove the 'problem' individual from the company.

In countries such as Spain, Italy, Portugal and Greece, retailers also reported that they were inhibited in dealing with thieves by national criminal codes under which shop theft by a customer did not become a crime until some minimum value was reached (usually between €20 and €30). It was reported also that in many cities in southern Europe the municipal police (who would normally deal with shop theft) refused to provide service where shop thieves were apprehended for amounts of less than €50. Although this seemed to be a matter of public policy in southern Europe, retailers in France, the UK, Denmark, Finland and Norway also made similar complaints. Local policies by municipal police forces meant that, in practice, retailers often had to ration their use of the police service—for example, they might only report persons who had stolen goods of more than a certain amount or who had previously offended. No such concerns were received from retailers in Germany or Austria.

Security and loss prevention costs

Throughout Europe, spending on security and loss prevention was an average of 0.29 per cent of retail turnover, or €5.781 billion. Security costs naturally varied between retailers. Different kinds of business had different crime problems and spending levels were affected by the vertical markets in which they operated. Retailers operating large stores such as hypermarkets, discount warehouses and, to a lesser extent, video, music or 'do it yourself' (DIY) stores, spent considerable sums on loss prevention and used a range of security staff and security equipment. In contrast, smaller stores selling clothing or shoes spent comparatively little on security: such stores had invested some money in electronic article surveillance and CCTV, and uniformed guards were often used in city centres, but the comments provided by companies indicated that they did not think it would be economic to provide high levels of security. Security costs were lowest in other non-food retailers (where shrinkage was highest) and in shoe and leather stores, and highest in department and general stores and in supermarkets and hypermarkets.

Figure 2. Components of European retail security spending, 2001

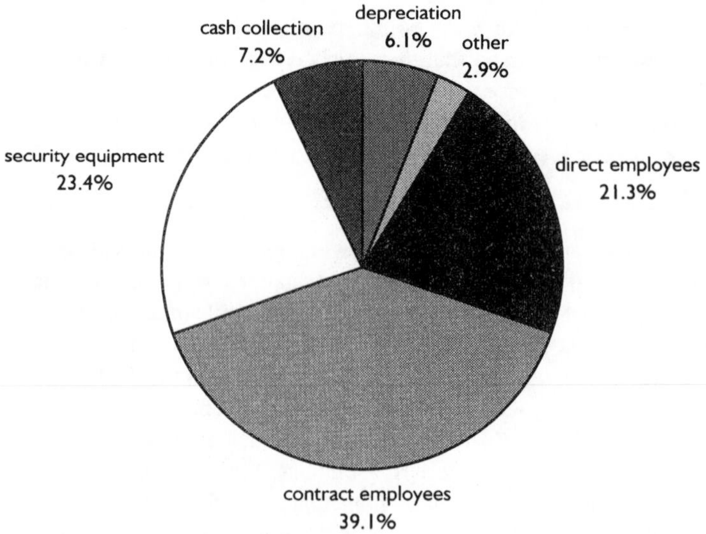

Of the spending on security in 2001, contract guards and store detectives accounted for an average of 60.4 per cent of the total budget, and of this percentage, two-thirds was allocated for contract security and one-third for in-house direct employees (Figure 2).

New capital investment in security represented 23.4 per cent of security costs, although the significance of these figures will be affected by the fact that some retailers will buy equipment outright and others will pay a lower annual amount to rent or lease the same equipment. The use of rental or leasing will apparently depress the rate of spending on security. A further 6.1 per cent of security spending went towards depreciation. The costs of cash collection varied considerably, but represented an average of 7.2 per cent of the total security budgets.

Table 7. Comparison of security spending in the UK, France and Germany

	UK	France	Germany
Direct security staff	17.1%	19.6%	27.7%
Contract security staff	31.8%	43.1%	30.9%
Combined staff spending	48.9%`	62.7%	58.6%
Security equipment	22.7%	17.3%	17.9%
Cash collection	18.5%	11.4%	7.4%
Depreciation	5.4%	7.4%	8.8%
Other	4.5%	1.2%	7.3%
	100.0%	100.0%	100.0%

A comparison of proportional security costs in the UK, France and Germany shows some clear differences (Table 7). France and Germany spent a much larger percentage of their loss prevention budget on security staff, a total of 62.7 and 58.6 per cent respectively, compared with the UK's average spending of 48.9 per cent. A large proportion of German security staff were directly employed in-house, whilst French retailers spent 2.2 times on contract security what they spent on in-house security. UK retailers were somewhat in the middle, spending on average 1.9 times on contract security what they spent on directly employed security staff. UK retailers tended to spend more on security equipment, mainly CCTV, than their continental counterparts, and also a considerable amount more on cash collection. The higher figures for depreciation in France and Germany may simply reflect the accounting conventions of those countries, or they may indicate that retailers in those countries were more likely to purchase crime prevention equipment than to rent or lease it. One conclusion that can be drawn, however, is that French and German retailers rely much more heavily on security staff to combat crime, while UK retailers may well spend disproportionately more on security equipment, mainly CCTV. Whilst it may be plausible to argue that this difference is responsible for the national shrinkage differences noted earlier, we do not yet know enough about the accounting conventions used, which may also play a part in these results.

Conclusions

This chapter indicates that it is possible to carry out a relatively small-scale security survey across Europe assisted by local publicity and local support to obtain a satisfactory amount of data on retail crime.

The results showed that although Europe is increasingly a single economic unit, there are considerable differences in national levels of shrinkage, the UK figure being particularly surprising in relation to historically high levels of loss prevention spending and a wide range of retail-sponsored local crime reduction initiatives. The average rate of shrinkage was 1.42 per cent, 82.4 per cent of which was regarded as being crime-related, involving costs of €23.818 billion. The total cost of crime for the retail sector (crime-related plus retail security costs) was estimated to be €29.599 billion, equivalent to 1.46 per cent of retail turnover.

Although European retailers were showing increased concern about the cost of theft by employees, they regarded customer theft as their biggest problem. Customer theft was thought to account for an average of 45.7 per cent of shrinkage and staff theft for 28.5 per cent. The number of thieves apprehended by stores was 1.266 million, 94.9 per cent of which were customer thieves. Apprehended staff stole on average thirteen times more in value than the average apprehended customer thief. Larger retailers, particularly those operating large stores, made a robust response to retail crime, spending a total of €5.781 (equivalent to 25 per cent of total crime-related losses) on retail security.

In the face of many differences in spending and performance, there were three main areas for which retail responses were practically identical. Retailers reported that the same types of expensive, highly promoted, branded products were the target of thieves, irrespective of country. Many retailers also expressed serious concerns about the ineffective performance of their criminal justice system in providing worthwhile deterrents against offenders. Lack of police service to arrest and process customer thieves, particularly where relatively small values were involved, was also a common problem, undermining the retail loss prevention approach.

The purpose of this study was to provide retail managers with comparable data on retail theft and loss prevention, from a wide number of European countries, to help them undertake their role of managing security and loss prevention. The next stage, and certainly the more interesting topic, is to understand why so many of these figures differ.

Appendix. Retail companies which responded to the survey

Country	No. of companies which responded	No. of stores	Combined turnover (€ billion)
Austria	15	270	1.203
Belgium/Luxembourg	24	535	8.57
Denmark	18	238	6.075
Finland	18	939	6.616
France	56	6309	111.236
Germany	52	3480	61.616
Greece	8	480	5.271
Ireland	16	160	3.556
Italy	34	2176	48.746
The Netherlands	22	1958	17.874
Norway	15	375	3.686
Portugal	9	261	1.445
Spain	31	1302	14.13
Sweden	2	143	1.194
Switzerland	25	876	9.092
United Kingdom	79	10741	149.496
Totals	**424**	**30,243**	**449.806**

Notes

1 Joshua Bamfield is Director of the Centre for Retail Research, PO Box 5413, Nottingham, NG7 2BJ; email: research@retailresearch.org. The research for this chapter was supported financially and practically by an independent research grant from Checkpoint Systems Europe, which is gratefully acknowledged.

2 Notably the series from Hollinger, at the Security Research Project, University of Florida, the most recent publication being Hollinger, R.C. and Davis, J.L. (2002) *2001 National Retail Security Survey: Final Report.* Gainesville, FL: Security Research Project, University of Florida; and the series from the British Retail Consortium, the most recent being British Retail Consortium (2002) *9th Retail Crime Survey 2001*. Norwich: Stationery Office.

3 The series from EHI, now in its third year, is a welcome addition to a sparse literature. Horst, F. (2001) *Inventurdifferenzen 2001: Ergebnisse einer Aktuellen Erhebung*. Cologne: Verlag EuroHandelsInstitut. Apart from Germany and the UK, most countries have made no recent attempt to measure retail crime losses: for example, the last French survey was 1995, while Italy and Ireland have never carried out a survey of this kind.

4 Exceptions to this being Beck, A. and Bilby, C. (2001) *Shrinkage in Europe: A Survey of Stock Loss in the Fast Moving Consumer Goods Sector*. ECR Europe Shrinkage project. Brussels: Efficient Consumer Response; and Bamfield, J.A.N. and Hollinger, R.C. (2001) Managing Losses in the Retail Store: A Comparison of Loss Prevention Activity in the United States and Great Britain. In Mars, G. (ed.) *Occupational Crime*. International Library of Criminology, Criminal Justice & Penology. Aldershot: Ashgate.

5 Bamfield, J. (2002) *European Retail Theft Barometer 2001: Monitoring the Costs of Shrinkage and Crime for Europe's Retailers. A Report to the Retail Industry 1.* Nottingham: Centre for Retail Research.

6 For example, Bamfield, J. (1992) *National Survey of Retail Theft and Security: Final Report.* Northampton: Nene College (this survey also collected data from smaller retailers); Burrows, J., Speed, M. and Bamfield, J. (1995) *Retail Crime Costs, 1993/4 Survey: The Impact of Crime and the Retail Response*. London: British Retail Consortium.

7 British Retail Consortium (2000) *Retail Crime Survey 1999*. London: BRC. The most recent BRC survey has suppressed some data necessary for this calculation.

8 Surveys using this approach for retail crime, which in view of its high cost is usually only feasible where public funding is involved, include Mirrlees-Black, C. and Ross, A. (1995) *Crime Against Retail and Manufacturing Premises: Findings from the 1994 Commercial Victimisation Survey.* Research Study No. 146. London: Home Office; Burrows, J., Anderson, S., Bamfield, J., Hopkins, M. and Ingram, D. (1999) *Counting the Cost: Crime Against Business in Scotland.* Edinburgh: Central Research Unit, Scottish Executive Home Department Central Research Unit, with Scottish Business Crime Centre.

9 The US definition of one billion (= 1000 million) is used in this chapter following the conventions of UK economics and business. The term 'turnover' is used to mean 'revenues' throughout Europe, and consists of sales plus total value added tax. At the time of this survey, €1 was equivalent to £0.62 sterling.

10 British Retail Consortium, op cit, p 47.

11 Hollinger and Davis, op cit.

12 Beck and Bilby, op cit, p 12.

13 Horst, op cit, p 20.

14 Hollinger and Davis, op cit, pp 3–4.

15 At a press conference held to present the results of this survey in Frankfurt in January 2002, journalists (usually the most cynical of persons) from the German business press and daily newspapers seemed amazed that retail companies might think that more than a handful of staff would ever steal from their employers.

Chapter 9

Shoplifting: Patterns of Offending Among Persistent Burglars

Jacqueline L. Schneider[1]

Conspicuously absent from the literature is a thorough investigation into the role that shoplifting plays in the lives of persistent burglars. The findings presented in this chapter are part of a large-scale research project on stolen goods markets in Shropshire, England.[2] *Previous research has suggested that shoplifting helped fuel the stolen goods markets; therefore the project aimed to target shoplifting as part of its strategic initiative. Shoplifting, among this sample, is shown to be largely a rational choice among persistent burglars. Additionally, such burglars appear to be shoplifting regularly as a way to earn extra money and to obtain property that is not available during burglaries. Patterns of shoplifting and interview findings on drug use are described. Finally, suggestions for crime prevention and reduction are offered as a way for managers to develop security solutions.*

Introduction

This chapter presents findings from a segment of a large-scale study which examined how stolen goods markets operate within the Shrewsbury policing division of Shropshire. From April 2001 to May 2002, West Mercia Constabulary implemented the 'market reduction approach' (MRA),[3] an innovative crime reduction strategy that aimed to make it more risky for thieves to steal, and subsequently to sell the stolen property. While any acquisitive crime can provide the goods that are bought and sold illegally, burglary and car crime remain at the centre of public and government attention in terms of prevention and reduction tactics. However, there is a void in the literature as to the extent to which prolific thieves, specifically burglars, engage in the less serious crime of shoplifting.

Therefore, the aim of the research was to see if prolific burglars steal regularly from shops, as preliminary research suggests.[4] There were three main objectives of the research. First, it sought to examine how frequently burglars engage in shoplifting. The second objective was to examine burglars' patterns of shoplifting. Finally, the research sought to identify strategies that industry and government agencies can adopt in order to prevent theft, as well as to reduce the motivation of thieves to steal. The findings presented here highlight the overall important role that shoplifting has in the careers of persistent burglars, and the need to elevate the status of shoplifting in terms of enforcement and prevention tactics.

Two rounds of face-to-face interviews were conducted with prolific thieves who were currently under sentence for burglary by courts in the Shrewsbury area. Study participants had to be 18 years of age or older and have a prolific history of engaging in acquisitive crimes. Other participants were considered if under sentence for other serious theft offences, but only if they had numerous and recent convictions for burglary. An official history of drug abuse/use was optional. Participation was strictly voluntary, with confidentiality and anonymity assured to those agreeing to be interviewed. The first round of interviews was conducted prior to any policing intervention taking place. Data from these interviews served as a 'baseline', describing general patterns of offending. The second round of interviews took place after a series of targeted police operations, and results were compared with the first round of responses. However, for the purposes of this chapter, findings will be presented for the entire sample.[5]

First, a brief description of the West Mercia project is provided in order to put the current research into context. Findings from qualitative interviews are outlined, focusing on the role shoplifting plays in the lives of persistent offenders in Shrewsbury, well as on their patterns of offending. The final section of the chapter discusses the strategies that shops, businesses and Crime and Disorder Partners can undertake in order to prevent shop theft, and consequently, burglary.

'We Don't Buy Crime': a background to the research

In 1998, the MRA was put forward as a crime reduction strategy which sought to disrupt stolen goods markets and therefore decrease theft by reducing the motivation of thieves to steal. The concept relies on the systematic and routine analyses of data for the purpose of guiding police operational tactics.[6] Of central importance to the concept are qualitative interviews with offenders.[7] By talking directly with those responsible for stealing, a clearer picture of the nature of thieving and the stolen goods markets can be developed. From this, strategic interventions can be developed so that the crimes that fuel these markets, as well as the markets themselves, can be reduced.

In 2001, West Mercia Constabulary was awarded funding under the Home Office's Crime Reduction Programme to implement the MRA. Under the rubric 'We Don't Buy Crime' (WDBC), the project used findings from previous research and specific

MRA data to shape police tactical operations.[8] Of specific importance to this segment of the research is the focus that West Mercia Constabulary placed on shoplifting. Historically, shoplifting is a crime that has not received a great deal of police attention or resources. While forces may implement specific initiatives that aim to reduce shoplifting,[9] police performance indicators, such as burglary, car crime or robbery, remain the priority over crimes that pose no immediate danger to the public. However, believing that shoplifting helped fuel stolen goods markets and that persistent and prolific burglars were committing it regularly, the WDBC project made a tactical decision to elevate shoplifting in terms of operational priority.

As previously mentioned, preliminary research on stolen goods markets revealed that shoplifting played a supporting role in providing supplemental income for prolific burglars.[10] Interviews were conducted with such burglars in Shrewsbury to see if this was the case. There were two primary reasons for the focus on shoplifting: first, property stolen from shops helps maintain stolen goods markets; and second, it might be easier to catch shoplifters than burglars. Therefore, if the preliminary findings were accurate, targeting prolific shoplifters might increase the odds of apprehending burglars, thus reducing burglary in the long run. The issue of rationality was also explored in order to assess the extent to which shop theft was a calculated decision among these burglars. If thieves are entering shops with a clear idea as to what they want or need to steal, shops can re-arrange displays that highlight targeted property. In other words, retailers would be in a better position to make decisions as to how to protect items they know are most at risk of being stolen—either through store design, camera or employee placement, or through other preventive measures.

The following is a description of the interview findings. General demographic characteristics of the sample are provided, and because drug use/abuse plays an important role in theft,[11] self-reported patterns of drug use/abuse are also included. The next section of this chapter explores the frequency with which burglars engage in shoplifting. Following this is a description of offending patterns. Finally, prevention tactics are presented for the retail industry and Crime and Disorder Partners to consider.

The research participants

A total of 50 men were interviewed, 25 of these at the beginning of the project, in May/June, 2001. After a series of targeted police initiatives, an additional 25 were interviewed in January/February, 2002. There was an average age of 22 for the entire sample. Both samples were almost evenly split between juvenile and adults, both pre- and post-implementation. A total of 27 juveniles were interviewed (13 pre- and 14 post-implementation), and 23 adults participated (12 pre- and 11 post-implementation).

Of those interviewed, 90 per cent (n = 45) had convictions prior to their current sentence. The total number of convictions for the sample was 730 (mean = 14.6) for 1668 offences (mean = 33.36).[12] The types of offence for which interviewees were convicted are shown in Table 1. Because of the purposive nature of the sample, it is no surprise that theft-related crimes top the list. By design, the sample was expected to have high conviction rates for these types of offences. Miscellaneous crimes include regulatory and traffic infractions. Examples of offences involving segments of the criminal justice system include revocation of bail and failure to appear in court. Property crime deals with destruction of property rather than the theft of it. Surprisingly, convictions for drugs were lower on the list than anticipated. The following section discusses in detail self-reported levels of drug use. Given the findings, drug convictions would have been expected to be higher.

Table 1. Interviewee convictions by offence type

Type of offence	Number	Mean
Theft	865	17.30
Miscellaneous	301	6.02
Police, courts, prisons	170	3.40
Property	82	1.64
Against persons	62	1.24
Cautions	60	1.20
Public disorder	56	1.12
Drugs	27	0.54
Fraud	26	0.52
Firearms	10	0.20
Non-recordable	9	0.18
Total	**1668**	**33.36**

Some of the interviewees (n = 20, 40 per cent) lived with both parents, while 34 per cent (n = 17) had lived only with their mothers while growing up. Single fathers, social services, and other living arrangements accounted for six per cent respectively. Two interviewees were cared for by social services. Of those who lived in single parent households, 38 per cent saw the other parent infrequently or never. Half of the sample had no formal qualifications, while 24 per cent had obtained GCSE, ten per cent NVQ, and 14 per cent some other type of training/certificate.

Drug use

Questions surrounding drug use were divided into two main sections: use six months prior to being imprisoned, and use at some time in their lives. Subtle changes in the responses were noted (see Table 2). However, changes were not statistically significant. Ninety-six per cent of the interview subjects admitted to taking drugs six months prior to their current incarceration, and 98 per cent to taking them at some point in their life. The majority of interviewees (n = 31, 62 per cent) stated that their drug use began prior to their involvement in crime.[13] The average age of taking their first illicit drug was 12.8, and 12.5 when they committed their first crime. However, the difference between these means is not statistically significant.

There was a large range in the reported average weekly cost of drugs, from less than £50 to over £2000. The majority, 66 per cent, spent under £400 each week on drugs. The main method of funding this was the commission of crime (n = 38, 76 per cent), while a further 16 per cent reported relying on a combination of drugs and legitimate employment. Shoplifting (n = 32, 64 per cent), burglary (n = 26, 52 per cent), other forms of theft (n = 34, 68 per cent), selling stolen goods (n = 44, 88 per cent), selling drugs (n = 24, 48 per cent), and other crime (n = 8, 16 per cent) were all reported as being committed in order to fund drug habits.

Table 2. Frequency of drug use

Type of drug	In six months prior to prison	At some time in life
Marijuana	74% daily use	84 % daily use
Amphetamine	14% a couple of times a month	52% daily to several times weekly
Ecstasy	34% daily to several times weekly	68% daily to several times weekly
Ketamine	90% never	4% several times a week
LSD	70% never	18% daily to several times weekly
Crack	4% daily to several times weekly	42% daily to several times weekly
Cocaine	22% daily to several times weekly	30% daily to several times weekly
Poppers	90% never	46% never
GHB	96% never	88% never
Methadone	12% daily	22% daily to several times weekly
Benzodiazepines	16% daily to several times weekly	26% daily to several times weekly
Alcohol	54% daily to several times weekly	66% daily to several times weekly
Heroin	40% daily to several times weekly	48% daily to several times weekly
Other drug	96% never	92% never

The range of crimes that are committed by thieves is alarming; so too is the frequency with which they commit them. Fifty-two per cent of interviewees (n = 26) admitted to committing crime on a daily basis, while an additional 22 per cent (n = 11) engaged in criminal activity 'several times a week'. A further 14 per cent (n = 7) engaged in crime once a week.

Shoplifting

Forty-four of the 50 interviewees (88 per cent) admitted stealing from shops. Additionally, 39 (86.7 per cent) of those who admitted shoplifting also admitted burglary. Shoplifting was the main source of income for 40 per cent of the sample. Burglary was the predominant way to make money for 42 per cent of those interviewed. Ten per cent said shoplifting was their main money-making activity for a short period of time, before they moved on to more lucrative types of theft, such as burglary or car crime.

Not surprisingly, shoplifting was the first crime that 26 per cent of the interviewees committed. The average age when shoplifting began was 12. Following closely behind is burglary (n = 11, 22 per cent), with the average age at onset being just under 13. The remaining crimes included car theft, other forms of theft, criminal damage, drugs, robbery, and use of firearms. Just under a third of the interviewees stated that the main reason they committed their first crime was because all their friends were doing it (n = 15), whereas 68 per cent said this was the reason for taking their first drug. Excitement, drugs,[14] provocation and other non-specific reasons were the next most frequently reported reasons.

Of those who said that shoplifting was the first crime they committed, 92.3 per cent (n = 12 of 13) went on to commit burglary and 100 per cent continued to shoplift. Approximately 82 per cent (n = 9 of 11) of those whose first crime was burglary admitted to shoplifting, whereas 100 per cent continued to burgle. While it is difficult to generalise findings from these data, there is a suggestion that a career that ends in burglary might well have begun with shoplifting. The possible progressive effect existing between these crimes warrants further research, as the two were certainly entwined in the present case.

When asked why they stole from shops, a range of responses were evident, including the fact: that goods were easy to sell (32 per cent); that they were equally easy both to steal and to sell (14 per cent); that there was always a demand for the goods (12 per cent); and simply that they were easy to steal (ten per cent). Of those that had shoplifted 33 (75 per cent) of the interviewees said they knew what they were going to steal prior to entering the shop, versus the 11 (25 per cent) who shoplifted but did not know what they wanted to take prior to going in. Unfortunately, due to sample size and the patterns of response, the data were not conducive to statistical analysis. However, they do show an interesting trend that has been previously overlooked in the literature—the shoplifting patterns of prolific burglars. The vast majority of the prolific, persistent burglars shoplifted *and* they knew what they were going to steal prior to entering a shop. From this, the rationality of their shoplifting behaviour can be deduced.

Prior knowledge of what to steal was due to the thieves either stealing to order or to their having a general idea of what general types of items could be sold regularly through the stolen goods markets. For example, the thief might not have been asked specifically to steal a DVD player, nor might he have the specific intent to steal one prior to entering the shop. However, if there was an opportunity safely to steal one, the thief would take it because he knew the probability of selling it successfully was high. Of those who admitted to committing burglary (n = 45), only 29 (65 per cent) interviewees said that when they shoplifted they knew what they were going to steal before entering the shop. An additional ten who shoplifted and also burgled said they had no idea what they were going to take prior to entering.

In order to highlight the meaning of shoplifting to prolific burglars, one interviewee stated: 'If someone comes up and asks me if I can get them a DVD player or a laptop computer, I'd go 'yeah'. And if on my next burglary there wasn't anything like that, I'd go shoplifting to get it.' Shoplifting was seen as a way to expand the inventory of the goods they needed or wanted to steal. In other words, instead of just relying on burglary as a way to obtain the necessary property, some would also turn to shoplifting in order to get items they needed to sell on. Shoplifting was also seen among some of those interviewed simply as a way to earn extra money. For example, one thief stated: 'If I need extra money, I shoplift.' Additionally, age seemed to help some thieves to become better at their trade. For example when asked how lucrative shoplifting was, one interviewee said: 'Make from shops? Not much to begin with, but when I got older, I got wiser. I'd make about £50 a week when I was younger and now £200 a week with just clothes.'

Discussion

This research is important for two primary reasons: first, it highlights the need for policy makers to elevate the importance of the crime of shoplifting; and second, it highlights the need to address what appear to be two divergent perspectives on the part of police and retailers with regard to shop theft.

The generalisability of these findings is limited, given the sample size and the purposive nature of its construction. However, the results call into question the notion that shoplifting is a crime generally committed by young people who pose little harm to society, or that it is a crime that offenders 'phase out of', or that it is a crime strictly bound by opportunity. This research shows that prolific and persistent burglars engage in shoplifting during their thieving careers, either as a way to supplement income, or as an alternative to committing burglary when it is difficult for them to do so. Data from this study also show that, more often than not, thieves know, in advance of entering the shop, what they intend to steal. Rational choice, coupled with opportunity, might prove helpful when trying to explain how and why shoplifting occurs among prolific thieves.

The thieves interviewed for this study were extremely clear as to what general types of property, as well as specific items, they intended to steal. They knew what they wanted or needed to steal, and if burglary did not yield the goods, shoplifting was the crime of next resort. The primary reason they were so organised in their thefts was the knowledge that there was a structure available for getting rid of the property—either because the goods were ordered or because they had handlers or members of the general public who were prepared to buy the items they stole.

As part of the problem-solving strategy, data were collected and analysed on the various crimes that contribute to stolen goods markets. Burglary and car crime were automatically included in the scheme as important and obvious contributors to such markets. These crimes are police performance indicators and therefore of major concern to the police. However, a tactical decision was made to focus on shoplifting—a crime that traditionally receives less attention from the police. It would have been remiss to ignore shoplifting as a supplier to the illegal markets just because it is seen as a less serious crime. Simply put, it is a crime that involves the theft of property, and that alone is enough to warrant inclusion. The unit of analysis for the MRA is property rather than the type and location of particular crime, or individual offenders.

While this chapter pays closer attention to the offending patterns of prolific burglars than it does to the explanation of stolen goods markets, the research is a first step in drawing attention to shoplifting in a wider context. The study situates shoplifting as an important contributor to a structured system of buying and selling stolen goods. By showing that chronic, serious thieves do engage in this crime, it shows that shoplifting is a mechanism by which thieves obtain demanded property in order to supply a hungry and consuming stolen goods market. Perhaps if shoplifting is seen in a more serious light, and as connected to other crimes that have far greater implications for victimisation, then resources and responses can be targeted accordingly.

Previous research on shop theft has questioned the notion that shoplifting was a 'function of rational choice'. Beck and Willis suggest that it might be unwise to 'categorise shop theft' as rational, which implies that the behaviour is merely one of opportunity.[15] They go on to say that if retail crime is unplanned, increasing costs to prevent these crimes might be wasted because preventing random crime is difficult, if not impossible. However, the present research counters this in terms of the behaviour of prolific, persistent thieves. While it is difficult to draw generalisations from the sample, it does provide a framework to suggest that chronic, persistent offenders might well be committing shoplifting with a significant amount of forethought. Given this, and coupled with the idea that these offenders might well be committing a disproportionate amount of the overall amount of shoplifting in the area, crime prevention techniques and efforts might go a long way towards reducing the prevalence of this crime, as well as others. Blanket statements such as those presented in terms of the irrationality of shoplifting might well be problematic when investigating prolific thieves. Furthermore, such statements might be dangerously misleading when trying to develop crime reduction strategies that aim to reduce other related crimes, such as burglary or stolen goods markets.

By tying in the seemingly rational aspect of shoplifting among some thieves with the fact that patterns of victimisation of shops can be monitored, crime prevention efforts can be fine-tuned so that policing can become more targeted and purposeful. For example, security managers can monitor what types of property are most frequently stolen. By reporting these trends to the police, a more detailed picture of the local demand for specific types of property could be drawn up. Once demand is recognised, shops can begin to 'design out' the risk of being victimised or take other preventive measures to protect the supply of the would-be stolen goods.

This raises another important point of the research—there appear to be two schools of thought about the policing of shop theft. Retailers believe they are doing their part to stop theft and that the police are lagging behind in their efforts to help combat the problem. The police, on the other hand, appear to believe that retailers are not doing enough, and that crime prevention advice comes second to profit. What is unclear is what the shops are doing to convince the police that a major problem does indeed exist.

Nationally, the retail community believes it is taking crime and its control seriously. Expenditure related to crime prevention in the UK was £0.6 billion in 2001 and £0.8 billion in 2000.[16] Businesses are hiring security staff and risk managers in order to protect merchandise and employees—for example, 50 per cent of the overall expenditure on crime prevention in 2001 went towards security staff. Other money is allocated to measures for securing cash (21 per cent); to protecting against theft (13 per cent); to protecting against burglary (seven per cent); to hardware leasing and maintenance costs (two per cent); and an additional seven per cent for other related costs. However, staff theft and customer theft account for the vast majority of retail crime losses, 35 and 47 per cent respectively. Therefore, a major question is why only 13 per cent of the crime prevention budget is allocated to protecting against theft.

Locally, there may be a different reality. Anecdotal evidence suggests that when small or nationally-owned retailers are approached with specific recommendations for prevention or reducing opportunities for crime, resistance can be encountered.[17] What becomes a contentious point is the need to make the retailing experience a pleasant one for the shoppers, versus the need to secure the items most at risk of being stolen. Retailers want their customers to have access to their goods, and some manufacturers do not like having their products behind screens or in cabinets. However, allowing customers greater access to merchandise puts at-risk items in danger of being stolen. The potential problem arises if retailers are warned of the danger, but choose nonetheless to do nothing to protect such items.

Though there is an emerging acceptance that businesses (regardless of size) ought to become active participants in crime reduction and prevention efforts, the extent to which their responsibilities extend is far less clear. Calls for government to meet the input made by the British retail community are being made.[18] It is commonly held that the police are reluctant to respond to calls from retailers when the latter's security staff apprehend thieves.[19] Studies also show retailers hesitating to involve the police because this takes up too much staff time, a reluctance to prosecute certain social groups, low success rates in court, and penalties seen as not having a deterrent effect.[20] However, if retailers respond to shoplifting without involving the police, the link between shoplifting and other crimes is missed. What is also missed is the opportunity to reduce more serious crime by attending to less serious but related forms of crime.

Additionally, levels of satisfaction when police are called are generally low.[21] However, what is not clear from the research is why this is so—is it due to slow police response time? Or to the fact that crime prevention advice was given but not heeded, and the police therefore feel frustrated? Or is the level of satisfaction low because the police see shop theft as a low priority, and not included in their performance indicators? What is also absent from the literature are studies that explore how the police perceive the efforts that retailers make in protecting their shops, merchandise and employees. Clearly, answers to these and other related questions need to be investigated scientifically.

Crime prevention efforts will not be successful if shops mount them in isolation, nor will they be successful if retailers are excluded from tactical responses by the police. Small and large retailers, and other members of the business community, have a corporate social responsibility to participate actively in reducing and preventing crime. Government officials must work in co-operation with the business and retail sector to devise effective strategies that deter thieves from stealing specific types of property. Data analyses stemming from police databases, prison interviews and interviews with shop victims can better inform retailers and the overall business sector on what property is most at risk.

A long-term solution is for police authorities to work in conjunction with industry for the purpose of designing theft out of the product and out of the shop. For example, research has shown that wearing safety seat belts while riding in a car reduced the likelihood of harm in an accident. Government regulated the industry such that cars were designed with safety belts included as a standard feature. In other words, the industry was required to 'design out' the potential victimisation. Why should the manufacturers of cameras or personal audio equipment, for example, not be made to build their products with theft in mind? Recently, some mobile phone companies have agreed to incorporate mechanisms in the design of the phone that will disable it, rather than merely the sims card, if it is reported stolen. If research shows that ownership or use of a particular type of property is associated with victimisation, should it not be incumbent upon the manufacture to help minimise the likelihood of that victimisation?

By advancing technological assistance, demands on police agencies might be alleviated because the motivation of thieves to steal has been effectively designed out of the product. Thieves will not steal something unless there is a demand for the goods and they have a way to sell them. Demand will significantly decrease if the item no longer has value after the point of theft. Additionally, some thieves will not steal if the risk of getting caught outweighs the financial benefit of stealing and selling the item.

Other strategies that Crime and Disorder Partners can follow, in conjunction with the retail/business community, to further reduce theft and other related crimes include:

- developing a property tracking system for post-theft identification;
- using facial recognition systems for shops and shop watch alert schemes, so that known, persistent thieves can be identified quickly and shops notified immediately;
- raising the seriousness of shoplifting to an appropriately high level among the public, police, courts and retail communities, through extensive media campaigns, educational programmes and better research;
- pointing out the effects of participating, directly or indirectly, in stolen goods markets to small businesses and to those corporate organisations that control some of the High Street retailers;
- streamlining court appearances so that businesses do not lose their employees' time in attending court day after day;
- using notification schemes for when prolific, persistent thieves are released from prison, so local shops can be prepared; and

- using banning orders for persistent, prolific burglars/thieves.

Admittedly, some of these are more difficult to achieve than others. For example, getting decision makers within the criminal justice system to see the connection between shop theft and other forms of crime can be an arduous task. Likewise, bridging the gap between police and retailers might prove as difficult. Raising the status of a crime that is historically seen as non-important to one that requires attention will not be easy. It is one that requires key policy makers and retailers to work together so that barriers and perceptions can be changed. Media campaigns can be designed so that the connection between shop theft and other serious crimes can be recognised. Additionally, the effects of such crimes, including increases in High Street prices and in overall crime, can be highlighted in the campaign.

Using the principles of 'problem-oriented policing', meetings between police and retailers should be convened regularly to identify the problem at the local level. During these meetings, retailers and police can exchange information and design tactical and innovative responses to the conditions. The purpose of the meetings would be to develop a mutually beneficial relationship, to facilitate long-term working partnerships. Once both sides understand fully the extent of the problem and what is being done to tackle it, tactical interventions and plans of attack can be developed. By dealing with shop theft from a partnership perspective, old perceptions and attitudes might give way so that innovative solutions could be initiated. But in order to function effectively in local partnerships, retailers need direction, leadership and support at corporate levels to achieve *real* crime reduction at the local level.

Once a greater understanding is developed, more effective and holistic prevention and reduction techniques can be formulated. For example, if further research confirms the pattern of shoplifting among burglars, burglary reduction programmes might well benefit from highlighting shoplifting as part of the overall aim of the programme. By targeting shoplifters in a systematic and routine way, burglary levels might well decrease. The converse is also true. Shop theft reduction programmes that neglect burglary run the risk of massive (all too detectable) displacement into burglary, with disastrous consequences on national targets for reducing that crime.

Notes

1 Jacqueline L. Schneider is a Lecturer in Crime Prevention and Investigation at the Scarman Centre for the Study of Risk Management, University of Leicester; email: js195@le.ac.uk.

2 Funding for this research was received through West Mercia Constabulary's 'We Don't Buy Crime' project, funded by the Targeted Policing Initiative, part of the

government's Crime Reduction Programme. The views expressed in this chapter are those of the author, not necessarily of the Home Office or of West Mercia Constabulary, nor do they reflect government policy.

3 For a complete discussion, see Sutton, M. (1998) *Handling Stolen Goods and Theft: A Market Reduction Approach.* Home Office Research Study No. 178. London: Home Office.

4 See Sutton, M., Schneider, J.L. and Hetherington, S. (2001) *Tackling Theft with the Market Reduction Approach.* Crime Reduction Series, Paper No. 8. London: Policing and Reducing Crime Unit, Home Office.

5 Future analysis depends on pre- and post-intervention comparisons. However, the information presented here can be treated independently of these.

6 For a complete description and discussion of the various types of data collected and analysed as part of the MRA, see Sutton *et al*, op cit.

7 This is not necessarily for the purpose of evidence-gathering. Rather, general patterns of exchange might prove more beneficial in the long term than information on specific offences. Disrupting the stolen goods market requires that attention be paid to all participants in the process—those who steal, who buy and who use the property. Market-level tactics require that operations expand beyond simply arresting thieves.

8 In addition to the systematic approach to collecting new forms of data, WDBC ran an extensive media campaign that focused on educating the public about the social and economic consequences of the stolen goods trade. Finally, an educational package was developed in partnership with the local educational authority. The package included materials that teach students about the implications of buying stolen goods in order to 'get a bargain'.

9 Winners of the British Retail Consortium's 'Safer Shopping Award' include Birmingham's Retail Crime Operation and the Winchester Partnership.

10 See Sutton *et al,* op cit.

11 Gossop, M., Marsden, J. and Rolfe, A. (2000) Drug Misuse and Acquisitive Crime among Clients Recruited to the National Treatment Outcome Research Study (NTORS). *Criminal Behaviour and Mental Health.* Vol. 10, pp 10–20; Jarvis, G. and Parker, H. (1989) Young Heroin Users and Crime: How do the 'New Users' Finance Their Habits? *British Journal of Criminology.* Vol. 29, No. 2, pp 175–85; Bennett, T. (1998) *Drugs and Crime: The Results of Research on Drug Testing and Interviewing Arrestees.* Home Office Research and Statistics Directorate Study No. 183. London: Home Office; Bennett, T. (2000) *Drugs and Crime: The Results of the Second Developmental Stage of the NEW-ADAM Programme.* Home Office Research and Statistics Directorate Study No. 205. London: Home Office.

12 This includes pending prosecutions (n = 31).

13 Existing literature states that a causal relationship between drugs and crime cannot be assumed; however, the high cost of drugs is often the main reason why crimes are committed. Shover states that drug users often begin their criminal careers before they are addicted, and that while drug use may not cause crime, heavy

drug use does increase the prevalence of criminal activity. Findings from the present research supports this—with well over half of the sample stating drug use began prior to involvement with crime. For a complete discussion, see Shover, N. (1996) *Great Pretenders: Pursuits and Careers of Persistent Thieves*. Boulder, CO: Westview.

14 Only 10 per cent of the sample said that drugs were the reason they committed their first crime. However, as addiction progressed, crime was the primary way by which thieves obtained money for drugs.

15 See Beck, A. and Willis, A. (1998) Sales and Security: Striking the Balance. In Gill. M. (ed.) *Increasing the Risk for Offenders*. Vol. II, *Crime at Work,* p 103. See also Farrington, D.P., Bowen, S., Buckle, A., Burns-Howell, T., Burrows, J. and Speed, M. (1993) An Experiment on the Prevention of Shoplifting. In Clarke, R.V. (ed.) *Crime Prevention Studies*. Vol 1. Monsey, NY: Criminal Justice Press.

16 British Retail Consortium (2002) *9th Retail Crime Survey, 2001*. London: BRC, with Reliance Security.

17 During the implementation of WDBC, some retailers were contacted to see if they would move at-risk merchandise from positions of vulnerability. Almost none of the shops that were approached agreed to reposition the items. The main reason offered was that corporate offices had to approve such decisions. At the time of writing, police are preparing to take an action plan to the business and retail community, in order to develop formalised structures between the interested parties so that innovative crime prevention strategies can be enacted together. Informal conversations with these retailers, and with others in the country, suggest that 'head offices' do factor losses from crime into their overall profit-making equation; and also that, at times, the practicality of crime prevention can be lost in practice. Retailers rely more on alarms, CCTV and staff to prevent the occurrence of shop theft than on the environmental redesign of their shops. For example, they find that to move items from boxes used in displays or to move the displays to different parts of the shop proves to be more cumbersome than traditionally used security measures. It is important to note that these responses have not yet been substantiated through scientific study, and it is therefore not known if they are the norm. However, the anecdotal evidence does suggest that the relationships between police and retailers at the local level might need investigating. Corporate heads within the retail sector may need to give more consideration to the balance between marketing a product at the local level and minimising the potential for crime opportunities.

18 British Retail Consortium, op cit.

19 Centre for Retail Research (2002) *The National Survey of Retail Crime and Security*. Nottingham: Centre for Retail Research.

20 See for example Mirrlees-Black, C. and Ross, A. (1995) *Crime Against Retail and Manufacturing Premises: Findings from the 1994 Commercial Victimisation Survey*. Home Office Research Study No. 146. London: Home Office; Wood, J., Wheelwright, G. and Burrows, J. (1997) *Crime Against Small Business: Facing the Challenge—Findings of a Crime Survey Conducted in the Belgrave and West End Areas of Leicester*. Small Business and Crime Initiative. Swindon: Crime Concern.

21 Ekblom, P. (1986) *The Prevention of Shop Theft: An Approach Through Crime Analysis*. Crime Prevention Unit Paper No. 5. London: Home Office; Beck and Willis, op cit.

Chapter 10

Reducing Employee Dishonesty: In Search of the Right Strategy

Martin Speed[1]

Employee dishonesty is a complex problem, and it is difficult for retailers to find an appropriate management strategy to combat it. This study investigates how employee loss prevention could be better targeted. It focuses on a particular major multiple-outlet retailer based in the UK. It uses this company's records of detected employee offenders and a survey of the attitudes of a representative sample of all their employees to investigate the characteristics of offending and offenders. The analysis shows that the attitudes of offenders and the characteristics of their offending differ according to their age and experience. A management strategy is proposed that divides employees into four groups, based on age and length of service, and targets a different approach to each.

Introduction

The British Retail Consortium (BRC) Survey of Retail Crime estimates that in 2001 employee dishonesty cost the retail sector £640 million.[2] As with theft by customers, retailers detect relatively few actual thefts, and the majority of the value is based on estimates derived from how much stock is found missing after stock checks. The latest report does not separate out a figure for detected dishonesty among employees, but in the 1993/94 financial year £20 million was attributed to clearly identified and recorded incidents of staff dishonesty.[3]

Research by Mirlees-Black and Ross at the Home Office in 1995[4] established a similar figure, of £22 million, for losses from clearly identified incidents of staff dishonesty, though they did not attempt to estimate losses from undetected incidents.

Many factors can put the retail work force at particular risk of being tempted into dishonesty. Employees are somewhat younger than in other businesses, and are thus closer to the peak age for offending. Many retail employees are employed on a casual or part-time basis, and therefore have less commitment to their employer. Wages in retailing are relatively low, providing a possible cause of grievance and therefore a potential self-justification for dishonesty. In addition, the nature of the work, with responsibility for merchandise and large sums of money, means that opportunities for dishonesty are difficult to eradicate.

However, employee dishonesty is a complex phenomenon, less clear-cut than other forms of criminality. An example of this is the way that perpetrators view themselves. Employee dishonesty displays the characteristic (which Sutherland[5] identified when describing his concept of 'white collar' crime) that employees can commit acts which the law defines as criminal, but still uphold a high sense of their own worth, and of moral superiority. This is an essential reason for Clinnard and Quinney, in 1967, to identify 'occupational crime' as a separate type within their typology of crime. They observe that:

> A major characteristic of occupational crime is the way in which the offender conceives of himself. Since the offences take place in connection with a legitimate occupation and the offender generally regards himself as a respectable citizen, he does not regard himself as a criminal. At most he regards himself as a 'lawbreaker'.[6]

Control or support?

Despite the complexity of the phenomenon, in the present author's experience retailers rely on one of two strategies to prevent dishonesty among their employees. The first is a 'control' approach, which centres on removing opportunities for dishonesty and increasing the perceived risk of detection. The second is a 'support' approach, which intends to counter the motivation to be dishonest.[7]

The 'control' message stresses that offenders will be caught. It is concerned with increasing the risks of detection through better protection of potential targets and increased surveillance of employees, ensuring that the latter conform to standards and instructions put in place to ensure that they do not have an opportunity to steal. It is the strategy most commonly put forward by security specialists. Carson's 1977 manual, for example, asserts that 'a good security system removes temptations, or makes the temptations appear dangerous or unprofitable'.[8] The prevention message stresses the particularly unpleasant consequences for employees who steal or transgress the rules.

In contrast the 'support' message plays on more personal values, focusing on how staff members will be 'letting themselves down' and 'letting down the rest of the team' by stealing. The dire consequences of being caught stealing are also stressed, but not as something the company will gleefully impose, rather something offenders will bring on themselves by stealing, and something from which the company wants to protect them. The supportive message also stresses the benevolence of the company, both as a way of showing that the employee has really no need to steal, and also to emphasise the status, reputation and benefits the employee will be throwing away.

The difference between these two strategies is demonstrated by the different styles of training video produced by retailers to combat employee dishonesty. Some focus on control, aiming to scare potential offenders by telling them about the high risk of detection and the dire consequences that result from it. For example, one British video was called *You're Nicked*—the tone of this can be deduced from its title.[9] In contrast, other videos take a more supportive approach, where employees are warned of the unpleasant effect any dishonesty would have on their whole store team, and their responsibility to keep each other from temptation.

These two approaches have resonance with a similar divergence of approaches observed within general management by Douglas McGregor.[10] His theory 'X' (that people are by nature indolent, irresponsible and need to be led) has affinities with the 'control' message, whilst his 'Y' theory (that people have the capacity to assume responsibility and can be encouraged to share organisational goals) has affinities with the 'support' messages. There are also similarities with Blake and Mouton's[11] concept of management styles fitting into a grid formed from two scales of concern: one of concern for production, and the other of concern for people.

It is difficult to pursue both strategies with the same vigour. To caricature the situation, it is difficult to convey a caring approach, to make an employee feel trusted, appreciated and part of a wider team, if he/she works under constant CCTV surveillance, among locks, bolts and constant checks. Consequently, there is a long-standing argument over which approach is most effective. To both camps their own claim is self-evident. It seems self-evident to one camp that employees who are happy and satisfied, and feel supported and looked after by their employer, will not want to steal from them, and researchers such as Greenberg have established experimentally that the way employees are treated will indeed affect their honesty.[12] It seems equally self-evident to the other camp that if you bolt everything down and watch your employees like a hawk they will have no opportunity to take anything. Most of the measures identified in Traub's[13] review of corporate employee loss prevention strategies were of this latter type.

To investigate the value of these two approaches a study was undertaken in one of the UK's larger retail businesses.[14] The study was based on two sources of data. A survey was conducted to investigate the attitudes of the employees as a whole to dishonesty, and to investigate their opinion of deterrents. Also, the characteristics of offenders and offences were investigated using the company's records of those employees who had been dismissed for dishonesty.

The attitude survey

The attitude survey was sent to almost a third (31 per cent) of non-managerial employees working in retail outlets in the target company. (Those working in distribution centres and head office were excluded, because their opportunities for dishonesty varied significantly from those working in the shops.) The effective response rate (after correcting for labour turnover) was similar to other surveys of the workforce that the company had undertaken, at 25 per cent. The sample was stratified to ensure it was representative of the company by age, length of service, gender, and location in the country. Despite the low rate of response, it was similar for each of these groupings, suggesting there was no particular group that did not respond.

The survey instrument was a written form. To give a private environment in which to deal with the forms, and to emphasise confidentiality, it was sent directly to home addresses. A pre-paid envelope was included for respondents to return the form directly to a survey company for data entry.

Prevalence of dishonesty

It was thought that questions asking directly about personal honesty might be too intrusive to encourage a response. It was decided that staff would be more willing to write about their perceptions of what 'staff in general' do, so the survey asked about the employee's perception of the honesty of his or her colleagues. In order to investigate the differences between different age and length of service groups, respondents were asked to consider only employees in the same situation as themselves. (The phrase 'people in my situation' was annotated in the section header to indicate that it meant people of similar 'age, rank and length of service'.)

The survey asked to what extent the respondents found a number of statements matched their own views. The statements were all in the form 'people in my situation never...' The options were 'agree strongly', 'agree', 'disagree', 'disagree strongly' and 'don't know'.

In most categories the majority of respondents were able to 'agree strongly' with the statements, indicating that they were confident that people in their situation were never dishonest in any of these ways. A smaller number disagreed, indicating that they thought some people in their situation behaved dishonestly. The most common form of deviance identified in this way was taking more time off than they were entitled to (19 per cent disagreed that people in their situation never did this), then came giving discounts to friends (ten per cent), making personal use of company property (six per cent) and taking home surplus free gifts (seven per cent). The survey respondents reported a very low level of serious dishonesty, with only one per cent reporting that people in their situation took home products without authorisation or borrowed money.[15]

Comparison of average losses per detected offender and the value of stock going missing in the target company suggests that a much higher proportion of employees are involved in dishonesty.[16] Instead, the prevalence of cash and stock theft indicated by the survey is very similar to the detection rates for these offences. Perhaps those who knew of undetected dishonesty failed to report it, or failed to return the survey form (making the sample less representative of employees who had already turned to dishonesty).

To facilitate analysis, the elements that were most appropriately considered 'dishonest' were combined into a single measure. This combined measure included employees whose disagreement could be interpreted as knowing that people in their situation sometimes took cash or stock or gave unauthorised discounts.

Age

When looking for factors that may have an influence on involvement in dishonesty, age is a natural choice. Many researchers have commented specifically on the relationship between the age of employees and their level of dishonesty. It has been found that younger employees are less influenced by the severity or the certainty of punishment,[17] and that they are more likely to be involved in 'social' theft.[18] It has also been suggested that younger staff steal more than their older counterparts as a response to their lowly social position in the organisation.[19] It is also likely that employees will desist from dishonesty as they experience 'maturing' events in their lives, but young males are less likely to give up theft than young females.[20] On the other hand, older employees have been found to be involved in more serious thefts.[21]

The distribution of the respondents who reported that people in their situation undertook some form of dishonesty is given in Figure 1. As might be expected, it shows considerable variations between age groups, with a higher proportion of employees in the younger age groups disagreeing that people in their situation were always honest.

Figure 1. Employee perception of dishonesty, by age group

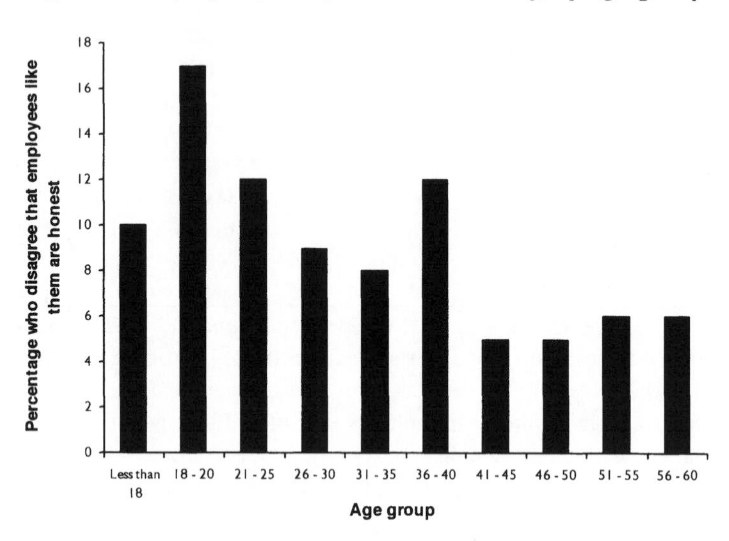

Length of service

It may also be anticipated that the length of service an employee has built up with a particular company will have an impact on dishonesty. Employees who have invested more of their time gaining status in a company are risking a greater loss if they are dismissed (though the factor is bound up with age, because most new starters are young, and it is impossible for youngsters to have built up long service). Figure 2 shows the results for the aggregate 'dishonesty' measure by length-of-service group. The groups most likely to disagree that people like them are always honest are those in the 1–5 and 6–10 year length-of-service groups. Very new starters and long-serving employees have the lowest proportions of disagreement with this view.

Attitude to deterrents

The survey asked respondents to what extent they agreed with a series of statements about factors influencing their personal honesty. Overall, employees reported their own standards as the most important factor, with 83 per cent strongly agreeing that these would influence them if they considered stealing. The possibility of a criminal record was the next strongest influence (79 per cent), followed by the anticipated effect on the respondent's family (76 per cent).

The thought of losing their job was ranked fourth (70 per cent). Given that losing their job is commonly the only sanction that results from being caught being dishonest, it is disappointing that it is ranked lower than other factors. What friends and workmates thought had the least impact (65 and 60 per cent respectively).

Figure 2. Employee perception of dishonesty, by length of service

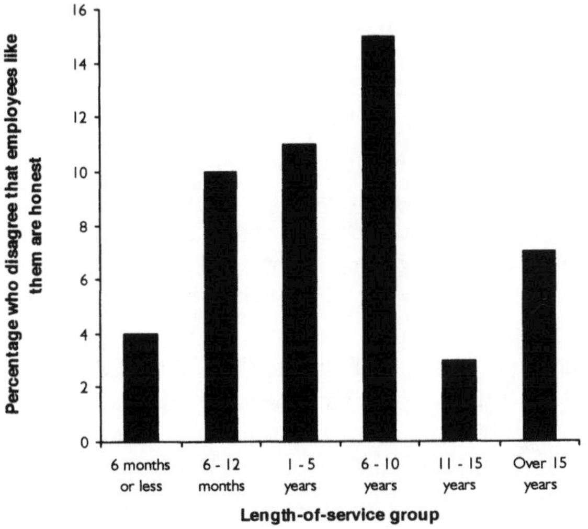

The impact of each factor is influenced by age. When broken down into four age groups (Figure 3), the factor with the highest impact reported in any age group is the 'effect on family' for employees over 41 years old (with 83 per cent of the group agreeing strongly that this factor would influence them).

Figure 3. Relationship between age and deterrent factors

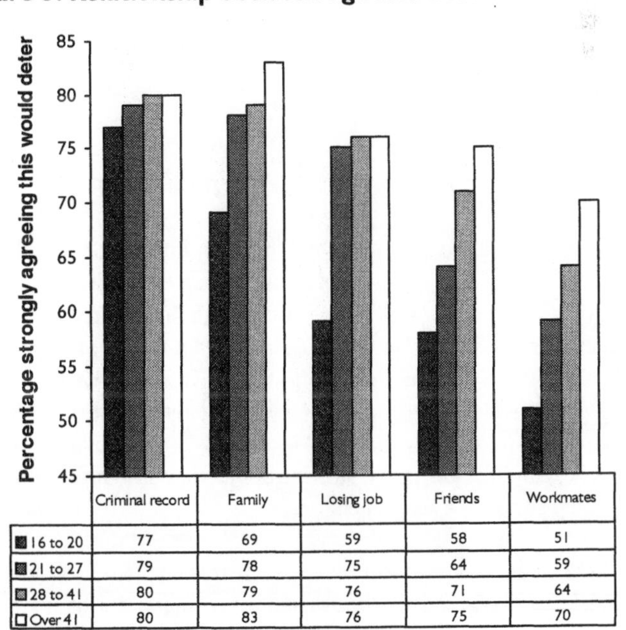

	Criminal record	Family	Losing job	Friends	Workmates
16 to 20	77	69	59	58	51
21 to 27	79	78	75	64	59
28 to 41	80	79	76	71	64
Over 41	80	83	76	75	70

Although the general trend is for younger staff to be less likely to be influenced by any of the factors, 'the thought of getting a criminal record' is a similarly strong influence on all age groups of respondents, with only three percentage points dividing the top and bottom of the scale.

'The effect dishonesty could have on my family' has the strongest influence on the oldest employees, but is also a strong influence on the 21–27 and 28–41 age groups. However, it is a much less strong influence on younger staff. This seems to suggest that employees are more affected by the impact their dishonesty might have on their family when they think of their children, rather than when they consider the effect on their parents.

Though less strong an influence than the effect on family, the thought of losing their job is reported as a similarly strong influence by all age groups except young staff. Employees in the 16–20 age group clearly do not believe they have as much invested in their careers as older staff. However, it is age that is important, and not the length of service in the company, because the lesser influence of 'losing my job' among the youngest age group is not mirrored in the lowest length-of-service group. Recent joiners (those with six months' service or less) generally agreed strongly that losing their job *would* influence them—71 per cent in this length-of-service group reporting strong agreement. Older employees clearly value their job more, even if they have just started it.

'What friends would think' and 'what work-mates would think' were factors that showed similar results to those for 'losing my job'. The younger the age group, the lower the proportion that reported they would be strongly influenced by these factors.

Perception of the risk of detection

Before any of these deterrent factors can have an impact, the employee needs to be caught. The risk of being caught is going to be a primary deterrent. Relatively few employees believed they could get away with dishonesty (ie gave 'disagree' or 'disagree strongly' in response to statements that they would be caught if they took one of the three actions in the question). Even for the least risky type of dishonesty proposed, 'taking home stock without authorisation', only 11 per cent of the respondents thought that they could get away with it. Seven per cent disagreed they would be caught if they gave stock to someone without full payment, and six per cent if they stole cash from the store.

If these three questions are combined, then this identifies 15 per cent of the respondents who disagreed that they would be caught for at least one of these types of dishonesty. Comparing this new measure with broad categories of age (as shown in Figure 4 below) shows that the perception of the risk of getting caught also varies with age.

Figure 4. Relationship between age and perceived risk of being caught for dishonesty

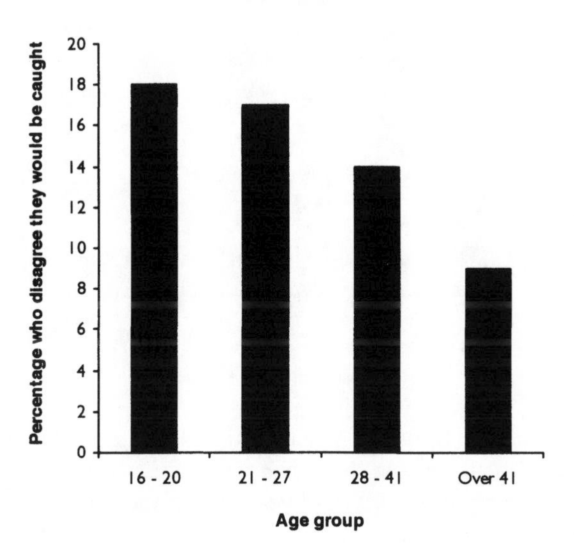

A higher proportion of younger staff believed that they would not be caught if they were to behave dishonestly.

However, a different impression is given by the figures on length of service. Figure 5 shows the percentages who disagreed they would be caught broken down into length-of-service groupings. The confidence in being able to evade systems among youngsters seems to be a product of age, not their length of time with the company.

It seems that new starters are least confident of avoiding being caught (eight per cent disagreed they would be caught), but employees who have been with the company longer, and learnt more about the way the company works, are more likely to think that they would be able to get away with dishonesty. The length-of-service group with the highest percentage thinking they could avoid being caught is the one containing those with 1–5 years' service (18 per cent). The percentage then reduces progressively for those with 6–10 years' service and the 11–15 years' service group. However, the small number of employees with over 15 years' service (those who probably have the best knowledge of company systems) is the group most confident they could escape detection.

Figure 5. Relationship between length of service and perceived risk of being caught for dishonesty

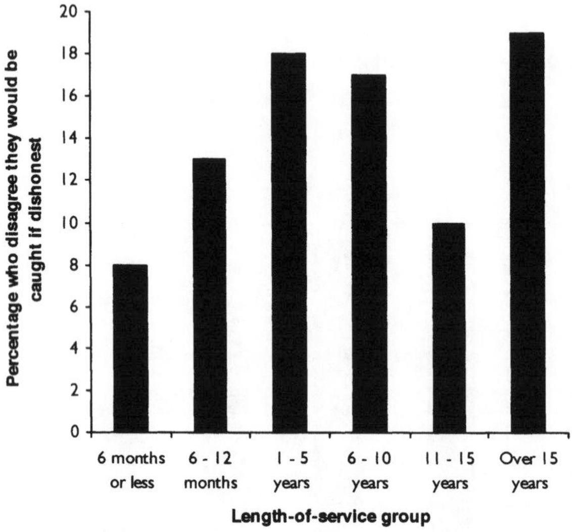

Length-of-service group	Percentage who disagree they would be caught if dishonest
6 months or less	8
6 - 12 months	13
1 - 5 years	18
6 - 10 years	17
11 - 15 years	10
Over 15 years	19

The group least likely to be deterred by the risk of getting caught, then, are those who, although relatively young, have had time to establish themselves in the company. Of those employees with 1–5 years' service who were also aged under 21, nearly a quarter (23 per cent) disagreed that they would get caught.

Resentment

Other studies[22] have established in the laboratory and through self-report studies that there is a relationship between dishonesty and resentment. The survey investigated this relationship.

It has to be recognised that this survey cannot be as effective a test of the relationship between dishonesty and job satisfaction as Greenberg's or Hollinger and Clark's work. This is because the dependent variable used here is not as clear-cut as laboratory observation, or self-reported dishonesty. The variable used here is whether the respondents attributed dishonesty to others in the same position as themselves.

The correlations between responses on job satisfaction and dishonesty are given in Table 1. Accepting the 'health warning' just mentioned, it can be noted that there are weak but significant positive correlations between the job satisfaction responses and the honesty responses.

The strongest correlation in this group is between 'getting on well with other staff' and 'not taking products home without authorisation' (0.28). The most consistent correlations are for the 'I enjoy my job' variable (over 0.20 in every case).

The correlation suggests that dissatisfaction and dishonesty are not unrelated. However, the importance of the finding will depend on how much dissatisfaction exists in the workforce. Here the findings gave some conflicting impressions. On the positive side, only one per cent of all employees felt that they did not get on with other staff members, and 92 per cent agreed they enjoyed their job. However, 24 per cent did not feel well treated by the company, and nearly half felt they were not always fairly rewarded for the hours they put in.

Table 1. Correlation between responses on job satisfaction and honesty

People in my situation:	I enjoy my job	I'm well treated by the company	I'm well treated by my manager	I get on well with other staff	I'm always rewarded fairly for hours I put in
Never make personal use of company property	0.210	0.148	0.192	0.211	0.181
Never take surplus free gifts	0.200	0.167	0.194	0.229	0.178
Never take home products without authority	0.222	0.193	0.203	0.279	0.201
Never borrow money from the company	0.223	0.174	0.186	0.256	0.161

Note: Spearman's rho. All these significant at the 0.01 level (2-tailed)

Dissatisfaction (which could lead to dishonesty) was also found to have a relationship with age and length of service. Figure 6 breaks down the percentage who disagreed with the statement 'I'm always rewarded fairly for the hours I put in' into length-of-service groupings. (This particular question was used because although it does not have the highest correlation with reporting dishonesty among people in the same situation, it was shown to be the most prevalent type of dissatisfaction.)

Figure 6. Relationship between sometimes not feeling fairly rewarded and length of service

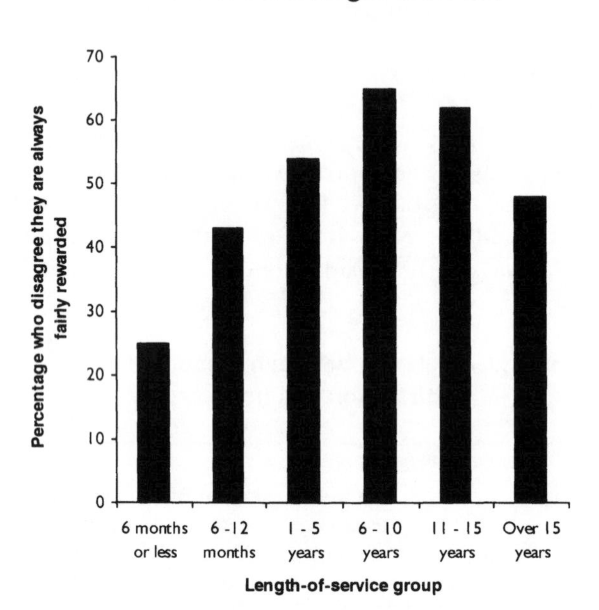

As was found by Hollinger and Clark,[23] it takes time to become dissatisfied, and the contracts given to new starters were generally seen by them as equitable. Those with less than six months' service had the lowest proportion (25 per cent) who sometimes felt unrewarded. For those with another six months' service the proportion was significantly higher, at 43 per cent. The length-of-service group with the highest proportion who did not agree they were always fairly rewarded for the work they put in was the 6–10 years group.

A similar pattern was found when relating this measure of dissatisfaction with age. Those in the youngest age group were less likely to feel unrewarded than those who were in the 21–27 or 28–41 age groups.

Summary of the survey findings

In summary, the survey shows that there are differences in the attitudes to dishonesty and in the experience of dishonesty shared by younger employees with less experience in the company and those shared by older, more established employees.

Younger employees thought dishonesty was more prevalent among employees in a similar position to themselves. If they were to consider acting dishonestly, fewer of the younger employees reported feeling they would be influenced by the thought of losing their job or of the effect on their family, or by what workmates or friends would think of them, than their older counterparts did. In addition, more of the younger employees thought they would get away with it if they were to act dishonestly, though a lower proportion of new starters believed they would do so, and it was the longest-serving employees who were most confident of all that they could escape detection if they wanted to.

The survey established a correlation between measures of dissatisfaction and measures of dishonesty within the workforce. However, the most common form of dissatisfaction (sometimes feeling unrewarded) was more prevalent among older, more established employees than it was among new starters.

Offender records

The company that formed the basis for this study keeps a database of all employees detected for dishonesty. These records contain information about the offenders such as their age, gender, and length of service, details of the offence, and details of the way in which each offence was detected. The records relating to two and a half years (from 8 February 1996 to 17 August 1998) were analysed as part of this study.

As might be expected from the attitude survey results, the offender records show a link between age and the prevalence of detection for dishonesty. The employees caught being dishonest are predominately young. In every thousand employees aged 18–20 years, 14 were found to be dishonest, whereas only two in every thousand were found to be dishonest in the 46–55 age group (see Figure 7).

Of course, 'being caught' is not a good basis for picking out a representative sample of dishonest employees, and the higher proportion of young staff being detected does not necessarily prove that more young staff are dishonest. For example, young employees have had less opportunity to learn how to avoid detection. They may be over-represented among detected offenders simply because it is easier to catch dishonest youngsters. However, the similarity to the survey findings suggests that the overall association of youth with increased offending can be trusted.

Length of service

Figure 8 shows the proportion of employees detected for dishonesty grouped by the length of service. The average for those with less than five years' service is one per cent of the workforce—but for those with six or more years' service, the average is less than half that. The general association of long service with less dishonesty is similar to the attitude survey results. However, the survey found that fewer new starters reported dishonesty among people in their situation than almost any other group. The inexperience of new starters will probably make them more vulnerable to detection than most other dishonest staff, and this has probably increased their representation in the detection figures. However, with longer service employees are more likely to have heard of dishonesty among other employees, boosting the levels of longer-serving employees reporting people in their situation offending. Therefore, the real level of offending probably lies somewhere between the two contrasting indications.

Figure 7. Detected dishonesty by age group

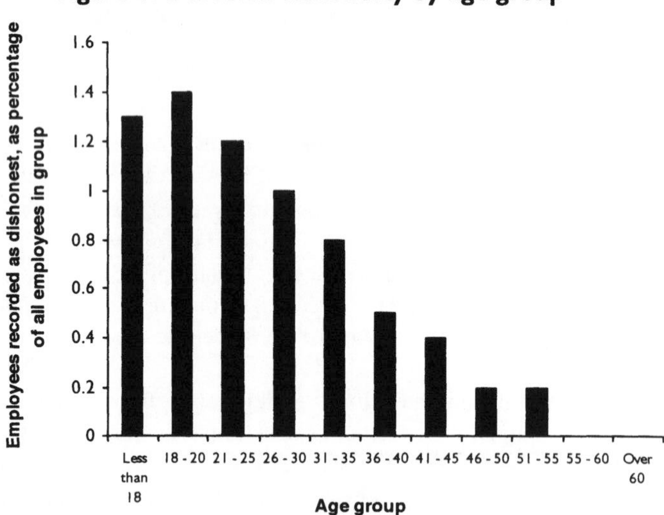

Figure 8. Detected dishonesty by length of service

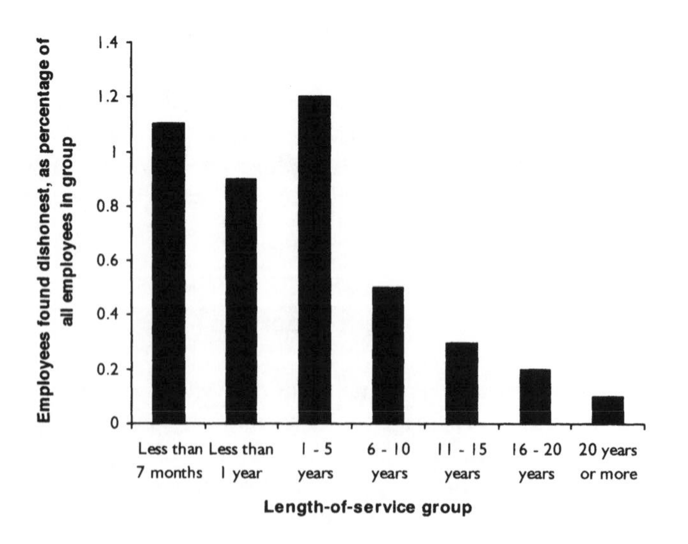

Age and length of service

The relationship between age and length of service was investigated by determining detection rates for employees grouped by length of service and then split into 'young' and 'old' categories. (The category 'old' meant age 26 or older.) It was found that the younger new starters were much more likely to be found dishonest than the older new starters. (For example only five in every thousand 'old' employees with less than seven months' service were caught, whilst 12 in every thousand under 26, with the same length of service, were detected.) Also, the older group has a considerably lower proportion of detected dishonesty in every length-of-service group. The younger group has a consistently high proportion of dishonesty. The peak is at a stage that represents quite long service for this age group: the 1–5 years' service group (with 16 in every thousand employees being found dishonest). The detection rate among those relatively few young employees with six year's service or more is lower (at nine in a thousand), but is still far higher than the older group with the same length of service (at five in a thousand).

Age, length of service and type of dishonesty

A particular advantage of using offender records is that quite specific details of the offence are recorded. These offences were grouped into general categories of similar offences. 'Refunds' was the category used to group all those incidents in which offenders took cash from the till and fraudulently recorded it as having been returned to a customer. A category of 'fiddles' was used to group together other general frauds based on the till system. A 'cash' category recorded simple thefts of cash, and the 'credit' category included all the frauds where the perpetrators benefited from fraudulent new applications for credit, or plastic card business. The category of 'collusion' was set up for incidents where the dishonest employee colluded with someone outside the business (usually by knowingly taking fraudulent credit cards). The category 'walk out' included all the simple thefts of stock from shops by employees, and the 'rubbish drop' category included those thefts of stock where the stock was put out with the rubbish for collection later. Finally, instances of copying software for profit were grouped under the 'software' category.

It was found that there was a link between the type of dishonesty detected and the age and length of service of the offender. Figure 9 shows the mean age and length of service for the types of dishonesty, summarised into the eight categories and plotted as a scattergraph. Aside from 'rubbish drop' thefts (which have a very low average age) and pirating software (which has a high average age) higher mean ages relate to higher mean length of service for each category of offence.

Figure 9. Mean age and length of service for each type of dishonesty

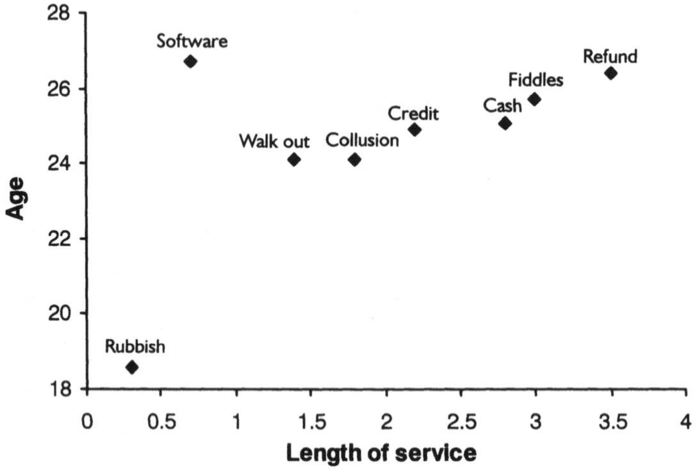

The offences rank themselves along a line of increasing age and length of service on a scale that could be described as representing increasing sophistication. At the bottom end, with an average employee age of 24 and length of service of one and a half years, is simple 'walk out' theft. At the top, with an average employee age of 26 and length of service of three and a half years, is refund fraud.

The apparent increase in the sophistication of the offence with age and length of service probably in part reflects the different opportunities open to employees with greater experience because of the increased responsibilities given them, and in part the greater skill in understanding company systems required to perpetrate them. Also, the least sophisticated crimes, with the lower age and length-of-service means ('rubbish drop' and 'walk out'), target stock, whilst the more sophisticated crimes with the higher mean age and length of service target cash. This may also reflect a preference associated with greater age and experience.

Being caught

How the offenders were caught is another factor that was found to show distinctions between groups of employees. There are two aspects to the process of getting caught, which were coded separately. The first is how someone in authority became alerted to the fact that there was dishonesty to investigate, and the second is how this particular individual was identified as being dishonest. Both aspects were found to be related to age and experience.

The mean age and length of service increases as the 'first alert' becomes more sophisticated. This is similar to the way that age and length of service were found to vary according to the type of offence—but this is hardly surprising as the two are so intimately related. Simple thefts of stock are detected by simple means such as searches, alarms, or finding stock missing or hidden. More sophisticated crimes, such as refund frauds, are detected by more experienced employees, such as area administrators.

It is similar for the second aspect, that of how an offence was attributed to a particular individual. It is the older and longer-serving employees whose dishonesty is hidden well enough to require a specialist investigator to link it to them; whilst employees with lower mean age and length of service were identified by non-specialist employees, or by being caught 'red-handed'.

A deduction from both aspects of the detection process is that the younger, less experienced employees have taken greater risks. They are caught red-handed in a search, or fail to take account of back-door alarms. The higher age and length of service of the employees whose dishonesty it takes a trained administrator to spot, and needs a specialist to investigate successfully, suggests that older and longer-serving employees take fewer risks.

Value of offences

It could be expected that employees with a greater length of service, having more to lose in terms of standing and status, would only risk dishonesty for larger amounts. Similarly, there is some logic in the notion that if younger employees are willing to take more risks, they may be prepared to do so for lower rewards.

The value of the offences associated with detected offenders suggests that this is the case. The offences associated with the young new starters (24 or younger, with less than two years' service) detected by the company averaged £214, whilst those associated with older, long-serving employees had a much higher average loss of £385. In line with expectations, on average the old new starters perpetrated higher-value offences than the young new starters. However, contrary to expectations, it was the young employees with two or more years' service whose offences had the highest average value, at £771.

In Table 2, the question of value is approached from the perspective of mean age and length of service for groups of values. This reinforces the notion that length of service rather than age is related to the value of the offence that employees are detected perpetrating. The mean age varies little between value groups (and is the same for both the £1–50 offences and the £501–1000 offences), whilst the average lengths of service are significantly higher for the higher-value offence groups.

Table 2. Mean value of offence cross-tabulated with age and length of service group

Value group	Mean age	Mean length of service
00. Not known	23.5	1.7
01. £1 to £50	24.7	1.9
02. £51 to £100	24.8	2.0
03. £101 to £500	23.0	1.4
04. £501 to £1000	24.7	4.1
05. £1, 001 and above	24.1	3.0

Summary of findings from offender records

The offender records show that young and newly started employees are more frequently caught for dishonesty. It was also found that the older and more experienced the employees, the more sophisticated the offence, and the harder it was to catch them.

In general it was found that older and longer-serving employees perpetrated more costly offences, and that length of service was a more significant factor than age. However, it is the combination of long service with low age that produced the highest average loss.

Conclusion: a targeted approach to prevention

There are two aspects of this research, when the study is considered as a basis for recommendations, where caution is in order. One concern is the low response rate to the survey, and the other is that it only dealt with one retailer, and thus one retail market sector. However, if the differences in the characteristics of offending and attitudes to offending among employees with varying lengths of service and in different age groups are representative of the wider retail workforce, then there are strong implications for the effectiveness of particular loss prevention strategies.

'Control' or 'support' strategies rely on different factors for their effectiveness. For example, a 'support' strategy will only be successful in persuading employees to consider the negative effect dishonesty can have on the whole team if the individual hearing the message is influenced by what colleagues think of him/her. Similarly, 'control' measures will only be effective with an employee who has not worked out ways around them. What this study suggests is that these factors vary with age and experience. The differences suggest that the question of identifying the most effective employee dishonesty prevention strategy can only be answered in the context of the age and length of service of the employees in question.

If the characteristics of offending and the impact of deterrents vary broadly with age and experience, then it seems appropriate that the deterrent strategy adopted by management should be varied to accommodate these changes. In a comparable way to how Paul Hersey's 'situational management' concept[24] develops Blake and Mouton's management grid, and targets different management approaches to employees at different stages in their career, it is probable that the management of staff dishonesty should differ for employees at different stages in their careers. There is considerable similarity between the measures in situational management theory of 'directive' and 'supportive' management behaviour, on the one hand, and the 'control' and 'supportive' approaches to employee dishonesty on the other. The degree of 'support' and 'control' could be varied according to the age and length of service of employees. This approach is represented diagrammatically in Figure 10.

Figure 10. Managing employee dishonesty

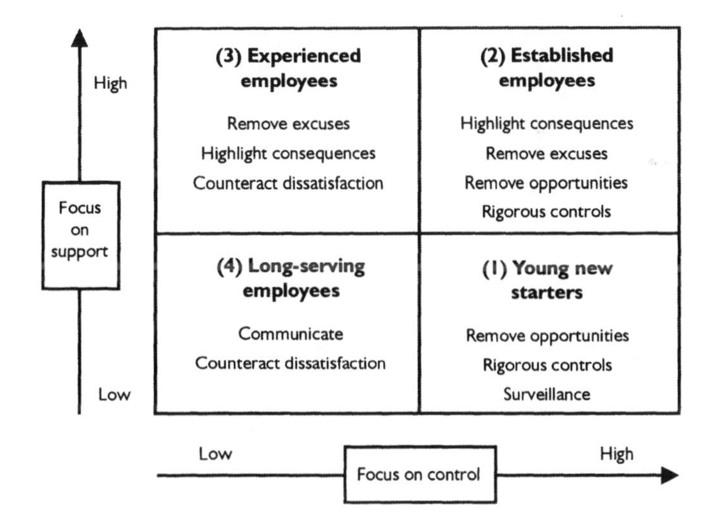

It is difficult to be precise about how to divide up the employees into groups, because age and length of service are so strongly related, but there appear to be four groups. First, there are the young new starters, around 20 years of age or younger, with less than a year's service. Then there are the young but established employees in their twenties who have been with the company for around two years. The third group represents experienced employees with longer service than this. The fourth group consists of employees who have considerably greater length of service, or are much older.

The evidence from this study is that the first group, the young new starters, are at greatest risk of being tempted towards dishonesty, and that they are less likely than other employees to be deterred from it by thinking of the effect their dishonesty may have on their family, friends or colleagues, or by the threat of losing their jobs. On the other hand, more of them think that they would be caught than do the slightly longer-serving employees. Many of the particularly young recent starters also do not seem as aware as older employees that dishonesty does take place in the company. The detection records suggest that they are more likely to commit the simplest types of offence, and to account for the lowest individual values.

For this group of employees a response based strongly on 'control' seems most appropriate, restricting access to high-risk operations and checking regularly that they are complying with systems. There seems less opportunity with these employees to persuade them that dishonesty is not in their best interests—it seems more effective to deny them opportunities.

The second group, established employees who are still relatively young but have gained experience of systems, present the greatest risk. Both offender records and survey reports put them at high risk of offending. They are more confident that they can avoid detection than are new starters, and these young but experienced employees were found to commit the highest value offences. However, compared to the young new starters they are more influenced by the potential consequences of detection, the risk of losing their jobs, and to some extent the effect on their family, friends and workmates.

This suggests that a strategy that is strong on both control and support is most appropriate to this group. A regime that reduces opportunities needs to be backed up effectively with a clear portrayal of the risks. A criminal conviction is the most persuasive factor to this group, so efforts to ensure that potential offences are portrayed as 'criminality' and not just 'rule-breaking' are likely to have an impact.

Offending among the third group, the experienced and older employees, is less common, but is more sophisticated, and less easy to detect when it does take place. It is this group who are most likely to feel unrewarded or dissatisfied with their treatment, and therefore at greatest risk of retaliatory criminality. However, these employees are more influenced by the effect of any dishonesty on their family, friends and colleagues than are younger employees.

Controls that remove opportunities are less likely to be effective with this group of employees. Their greater experience means that they are more likely to be able to bypass them. They are more likely to be influenced by being reminded of the status and benefits they have within the company, and by having the potential impact of offending on their financial security, social standing and family life portrayed clearly to them. A 'caring' approach with this group of employees is also needed to remove the potential for grievances, because that is likely to reduce the potential for retaliatory theft.

The fourth group, the small group of older employees with the longest service, seems to represent the lowest risk, and there is a case for expending the least resources on them. The number feeling unrewarded is lower, and they are the most influenced by all the deterrent factors investigated, particularly the effect dishonesty would have on their family.

There is very little that could be done to reduce dishonesty in this group by stricter controls. They were the group most confident that if they were to be dishonest they would not be caught, and given their experience of the controls in place, they are probably justified in thinking this. A 'supportive' approach is more appropriate. Apart from a greater proportion of this group being influenced by what their family might think of them, they are also more likely to be influenced by the impact their dishonesty would have on colleagues. However, they also reported less dissatisfaction, and after staying with the company longer probably do not need reminding of the status and benefits they have built up.

A limited, but more detailed, probe into the circumstances behind the detection records of some individual cases involving senior employees provided evidence that crimes representing a response to a severe personal problem that the employees could not share[25] were more common among this group. This finding, coupled with the less encouraging indicators for either a supportive or a control regime, suggests that good general personnel management practice (maintaining effective communication, identifying and countering grievances and personal problems, and building strong teams) seems more appropriate than specific crime prevention initiatives for this mature and long-serving group of employees.

Although this model runs contrary to a natural desire to treat all groups of employees in the same way, aspects of the approach are already in place. For example, some retail organisations set minimum ages or lengths of service for certain higher-risk operations, such as preparing cash for banking or operating the till. There are some excellent opportunities to introduce the approach. Induction programmes, for example, provide a natural platform for a dedicated employee dishonesty prevention message targeted at the characteristics of offending and attitudes to offending associated with new starters.

In practice, retailers vary most aspects of their strategy for human resources, such as bonus structures and training opportunities, between different sections of the workforce. Individual businesses may have to research the opinions of their own staff to ensure they target loss prevention accurately, but the message of this study is that the strategy cannot remain 'one size fits all'. Controlling employee dishonesty is not a question of finding the right strategy, but of matching the right strategy to the right group of employees.

Notes

1 Martin Speed is a crime information analyst working in the retail industry; email: martin@mcilroy-speed.fsnet.co.uk.

2 British Retail Consortium (2002) *Retail Crime Survey 2001*. London: BRC.

3 Speed, M., Burrows, J. and Bamfield, J. (1995) *Retail Crime Costs Survey 1993/4*. London: British Retail Consortium.

4 Mirlees-Black, C. and Ross, A. (1995) *Crime Against Retail and Manufacturing Premises: Findings from the 1994 Commercial Victimisation Survey*. Home Office Research Study No. 146. London: Research and Statistics Department, Home Office.

5 Sutherland, E. (1949) *White Collar Crime*. New York: Dryden.

6 Clinnard, M.B. and Quinney, R. (1967) *Criminal Behaviour Systems: A Typology*. New York: Holt, Rinehart and Winston, p 132.

7 Bamfield argues that perceived risk and motivation are the 'critical determinants of the level of staff theft'. See Bamfield, J. (1998) A Breach of Trust: Employee Collusion and Theft From Major Retailers. In Gill, M. (ed.) *Increasing the Risk for Offenders*. Vol. II, *Crime at Work*. Leicester: Perpetuity Press, pp 123–141.

8 Carson, C.R. (1977) *Managing Employee Honesty*. Los Angeles, CA: Security World Publishing.

9 The video *You're Nicked* was produced in the 1970s by the Dixons Stores Group.

10 McGregor, D. (1960) *The Human Side of Enterprise*. New York: McGraw-Hill.

11 Blake, R., and Mouton, J. (1964) *The Managerial Grid*. Houston, TX: Gulf Publishing.

12 Greenberg, J. (1990) Employee Theft as a Reaction to Underpayment Inequality: The Hidden Cost of Pay Cuts. *Journal of Applied Psychology*. Vol. 75, No. 5, pp 561–8.

13 Traub, S.H. (1996) Battling Employee Crime: A Review of Corporate Strategies and Programs. *Crime and Delinquency*. Vol. 42, No. 2, pp 244–256.

14 Speed, M. (2001) *A Targeted Approach to Managing Retail Employee Dishonesty*. Unpublished doctoral thesis, Surrey University.

15 This hierarchy is similar to the large-scale self-report study reported in Hollinger, R.C. and Clark, J.P. (1983a) *Theft by Employees in Work Organisations*. Lexington, MA: D.C. Heath, pp 34 ff.

16 Ibid, p 140.

17 Hollinger, R.C., and Clark, J.P. (1983b) Deterrence in the Workplace: Perceived Certainty, Perceived Severity, and Employee Theft. *Social Forces*. Vol. 62, No. 2, pp 398–418.

18 Hollinger, R.C., Slora, K.B. and Terris, W. (1992) Deviance in the Fast Food Restaurant: Correlates of Employee Theft, Altruism and Counter Productivity. *Deviant Behaviour*. Vol. 13, No. 2, pp 155–84.

19 Tucker, J. (1989) Employee Theft as Social Control. *Deviant Behaviour*. Vol. 10, No. 4, pp 319–34.

20 Graham, J. and Bowling, B. (1995) *Young People and Crime*. Home Office Research Study No. 145. London: Research and Planning Unit, Home Office.

21 Horvath, F. (1990) Self-Reported Work Place Theft, Use of Illicit Drugs and the Personal Characteristics of Job Applicants. *Security Journal*. Vol. 1, No. 4, pp 226–34.

22 For example: Greenberg, J. (1993) Stealing in the Name of Justice: Informational and Interpersonal Moderators of Theft Reactions to Underpayment Inequity. *Organizational Behaviour and Human Decision Processes*. Vol. 54, pp 81–103; Hollinger, R.C., and Clark, J.P. (1982) Employee Deviance: A Response to the Perceived Quality of the Work Experience. *Work and Occupations*. Vol. 9, pp 97–114.

23 Ibid.

24 Hersey, P. (1984) *The Situational Leader*. Escondido, CA: Center for Leadership Studies.

25 The cases were found to resemble the phenomenon of 'an unshareable problem' reported in Cressey, D.R. (1953) *Other People's Money: A Study in the Social Psychology of Embezzlement*. Glencoe, IL: Free Press.

Chapter 11

'Troublemaker' and 'Nothing to Lose' Employee Offenders Identified from a Corporate Crime Data Sample

Nick J. Dodd[1]

Actual employee offender data has rarely been utilised in previous analyses of employees offending against their employer. In this chapter such a data source is used, and demonstrates that it has an important role to play in organisations to prevent and investigate employee crime. However, this is only achievable if the organisation in question knows how to harness the potential of the data and use it accordingly. The analysis of the data in this chapter provides support for previous research findings that employee offenders are virtually indistinguishable from their non-deviant colleagues. However, it does establish that there are different types of employee offender within this case study who adopt very different approaches to their deviance in the organisational setting. This has not been demonstrated before, and has implications for future research and for the development of preventative and investigative strategies within organisations.

Introduction

In the past, most analyses of crime against organisations by its employees have relied on the opinions and attitudes of the workforce. Researchers adopted the survey and interview approach, which was effectively the only route open to them to glean information about this criminal activity.[2] It would appear that few researchers were able to directly examine actual criminal activity and its perpetrators. Those who were able to[3] developed hypotheses which have rarely been statistically analysed, or which concentrated on very specific types of offence against the organisation, such as embezzlement and long firm fraud. An alternative, and rarely exploited, source of data in this context is the files on employee offenders which organisations maintain. These are often rich in details of the offender, the offence and the method of perpetration. This corporate crime data (CCD) is, by rights, a very sensitive source of criminal intelligence, but if appropriately analysed it can provide a strong basis for understanding criminal behaviour in this context. It is one of the contentions of this chapter that such an understanding can provide an ecologically valid foundation for crime prevention strategies within the organisational context.

A handful of researchers[4] have utilised CCD in the form of closed files held on apprehended offenders and maintained by organisations and their investigative operations as a valuable source of employee offending data. These offences were high-volume criminal activity committed against organisations by either customers or employees. Dodd was able to develop a system of early identification of fraudulent insurance claimants who had defrauded insurance companies in the UK by analysing the profiles of offenders, their offences and their offending behaviour. Robertson was able to provide feedback to a distribution company about vulnerable security environments and methods of delivery of items passing through the company's system, by analysing all the information contained in CCD files, including the personal information. He was thus able to examine the nature of the security environment and how the organisation facilitated it, in addition to the individual characteristics of the offenders. In these contexts CCD proved a rich source of information on all aspects of the offence, from the individual to the organisational context. Such analysis enabled the researchers first to focus on the aspects of criminal activity which organisations deemed to be criminal, and second to demonstrate that this approach can be directly fed back to organisations, in the form of strategic approaches to reduce and solve crime being committed against the organisations in question. This was achieved through the analysis of corporate crime data, and also by establishing a firm understanding of the organisational context through exposure to the organisation and its employees.

It is not only organisations that have failed to harness CCD's true potential. Research in this area has spanned the disciplines of psychology, sociology and anthropology, with the main focus coming from criminologists. Such a breadth of interdisciplinary research reflects the various implications of such deviant behaviour, as well as the various factors that correlate with it. However, in spite of this breadth of interest this chapter reveals that very little research has been able to employ actual CCD as its information source. Some researchers[5] have successfully embroiled themselves in criminal cultures. Their research has taken a participant observer approach, and resulted in a set of hypotheses that have undergone no statistical analysis. Whilst these sorts of studies have provided valuable accounts of a 'fiddling culture' in action, they do not provide organisations with an easily accessible methodology for assessing the risk of crime as part of a strategic approach to its reduction. Corporations rarely have the time, resources or capability to carry out such long-term research. Without at least this capability, it is difficult for an organisation to translate social science approaches and findings into a strategy from which they can benefit. Some researchers[6] have called for a stronger relationship between criminology and business in this context. This is a valid point, but it is inescapable that a large part of crime against organisations by employees has a psychological aspect, and the contribution that psychological research can make should not be underestimated.[7] Therefore, it is also the contention of this chapter that ignoring such a rich source of material as CCD limits the understanding of crime in this context. Simultaneously, it leaves organisations with very little intelligence that can be operationalised in terms of reducing and detecting crime committed against them.

The expectation is that CCD will be a rich source of information on the individual offender. Individual offender characteristics have been the focus of considerable US-based research. Such factors as gender,[8] age and tenure,[9] economic pressure,[10] status,[11] trust,[12] belonging[13] and personality traits[14] have all been examined in some depth. Hollinger and Clark's[15] work has been the most significant in establishing that employees of low age and tenure are more likely to commit crime against their employer than their older, longer-tenured colleagues. They were also amongst the first researchers to state that economic pressure was unlikely to be a cause of employee crime. Whilst the findings of this research are of great importance in establishing that there is an identifiable 'profile' to the employee offender, it is difficult to establish firmly what this identity is from the research findings available. The reliability and validity of personality traits, in particular, have been shown to be questionable due to a lack of consistency over time and to the temporary nature of some deviant activities.[16] Thus, another contention of this chapter is that employee offenders are indistinguishable from their non-deviant counterparts.

To summarise, this chapter contends the following:

- CCD files constitute an ecologically valid source of information on employee offending;
- CCD (in the form of closed files on employee offenders) will be a rich source of useful information; and
- the typical employee offender is indistinguishable from his/her non-deviant colleagues.

Data source

In order fully to test hypotheses in this area, access to a large volume of CCD was required. Following considerable negotiation this was achieved with a parcel delivery company based in the United Kingdom. The necessary assurances of confidentiality and anonymity were made, and for this reason the company will be known only as 'Deliverance' in this chapter. Deliverance has its own in-house investigation unit (IHIU) which investigates all crimes or suspicious activity reported to it. By this token this unit is more reactive than proactive.

Initial contact with Deliverance revealed a high volume of loss of parcels in its system, resulting in high compensation claims and payouts. Management attributed a large proportion of this loss to theft by employees (between 30 and 40 per cent). It transpires that this was not supported by this research (see the 'Discussion and conclusion' section below).

The sample

Trawling the IHIU files produced 103 files suitable for analysis. It is worthwhile noting at this point that to produce 103 files involved a trawl back over three-and-a-half years. Any offence was considered to be relevant, and therefore cases cover damage, theft, wilful delay, evasion of fees, fraud and embezzlement.

The problem of producing a representative sample plagues the researcher in this discipline. Invariably, there are reasons why a sample population will never represent the general population—this is the ultimate dilemma of social science research. In particular, an apprehended offender population is not a representative sample of all offenders. It only comes close to being a representative sample of those caught, and of their shortcomings in effectively perpetrating the offence. To some extent offender files only represent relevant details of the enquiry and not, comprehensively, the offender in question. Files vary dramatically in their content and quality. This is not to say that files do not contain useful information. They are a rich source, but the researcher has effectively to trawl someone else's trawl in attempting to standardise the content of a sample of files that are standard only in format in the first place. These are obstacles that are not easily negotiable. Without designing a procedure, or experiment, which will produce desired information in an approachable format the researcher must adopt alternative approaches of gleaning information or data from someone else's data. This is one of the difficulties of employing ecologically valid data. To neutralise these biases as much as possible an approach known as content analysis[17] was adopted.

Content analysis

A content dictionary was designed to capture the information captured in the IHIU CCD files. This acknowledged all the elements discussed in the introduction to this chapter but also allowed for the inclusion of other possibly important material contained in the files. In other words this process was one of theory building, with the ultimate aim of hypothesis testing.[18] It was not simply the counting of the number of occurrences of a particular action or piece of information. It was anticipated that the inclusion of the material not previously considered by previous researchers (at least not in published accounts) may have some importance and should not be excluded.

The content dictionary was produced, consisting of seven sections. This is not reported here as this research focuses on the variables in only one section of the dictionary—that of employee offender characteristics.

Findings

Analysis of the Deliverance employee offender, via CCD files, was both univariate and multivariate. Whilst the univariate analysis can provide some important descriptive statistics for the typical Deliverance employee offender, it does not account for some of the interrelationships between these variables. Multivariate analysis does allow for this, and by employing a methodology called 'facet theory'[19] it is possible to interpret the multivariate relationships between variables in a complex data set. This chapter focuses on the findings of the multivariate analysis.

Because of the focus on individual employee factors that have been reported as correlating with employee offending in the past, the variables from the CCD files regarding the actual perpetrators were subjected to multivariate analysis. Through an iterative process of analyses of various combinations of variables from the content dictionary, it was possible to produce a meaningful interpretation using SSA.[21] This is illustrated in Figure 1.

A sample of individual employee variables was analysed to establish if there was any structure illustrated in the interrelationships of the data gathered on these individuals. This would help establish if any of the individual factors reported as relating to employee crime in the past would also hold true for this sample. Because there was quite a high level of missing data it was only possible to test for these factors within the parameters of the data set produced from the content analysis of the offender files.

The variables included in this analysis were:

1. Offender under age 28
2. Without partner
3. Without children
4. Non-Deliverance employee
5. Under 3.5 years
6. Non-vehicle owner
7. Previous convictions
8. No confession
9. Previous IHIU attention
10. Other counterproductive behaviour
11. Financial difficulty

These variables were all included in the SSA because they were all seen to indicate a lack of responsibility, as well as of commitment, to the organisation. They have all been reported by previous researchers as being factors related to deviance and low commitment and responsibility in the make-up of employee offenders. Being young, not having children and not having a partner have been demonstrated to correlate highly with deviant employees. The reasoning behind this is that employees who have little to lose in terms of what they are committed to outside the organisation are less likely to have strong feelings of commitment to their own levels of lawfulness within the organisation. This is because if they are apprehended for crime against their employer they have little to lose outside of the organisation. These variables were selected on the basis that they shared a common order of low responsibility/commitment, and the list thus is not designed to represent all employee characteristics from this sample. This analysis is designed purely to gain understanding of the relationships of the variables relating to low responsibility/commitment. Similarly, the list is not exhaustive. The variables listed here have been identified by other researchers, and were included on this basis. The list of variables was also restricted by the absence of information in this context. For some of these variables there was a low frequency rate and where this information was missing it was recorded as absent. Undoubtedly, there is a richer picture to be gleaned from CCD if a complete data set is made available, but the information is currently only recorded to gain convictions or justify dismissals, where appropriate, and not for criminal intelligence analysis.

SSA produced a two-dimensional solution with a coefficient of alienation of 0.16 in 13 iterations. Figure 1 shows the projection of the two vectors of the two-dimensional space. In this figure each point represents one of the 11 variables listed above. Quite simply, the closer together any two variables are in the two-dimensional space the more likely that the other variable will also be present. By contrast, it is unlikely that when an offender has come to previous IHIU attention (variable 9) he/she will also be in financial difficulty (variable 11) because these two variables are on opposite sides of the plot. It should be remembered that the relationships between all the variables affect where they are plotted in the space. All the relationships between all eleven of the individual employee offender variables are combined to produce the SSA plot.

The SSA plot reveals two regions of importance in this context. On the left-hand side of the plot is the criminal/deviant region and on the right-hand side the low commitment/responsibility region. It is important to note that there are no variables in the top half of the plot, and this suggests that information about the whole research domain is missing from this small data set. This is unsurprising for a data set of only 11 variables drawn from CCD files.

Figure 1. SSA of individual employee variables

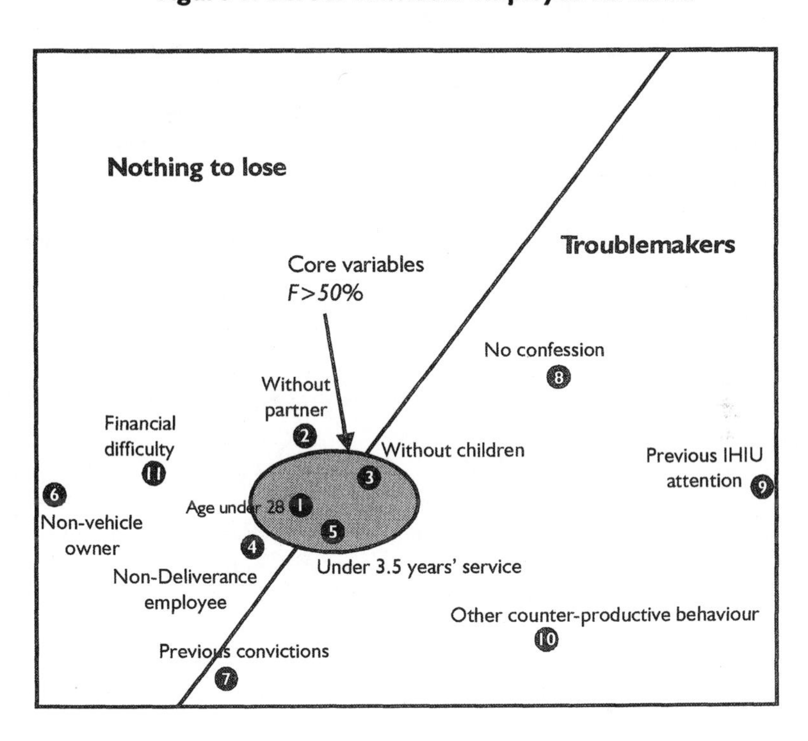

Troublemakers

This region of the SSA reveals that there is a deviant element to crimes committed against Deliverance. These are characterised by offenders who have previously come to IHIU attention, who do not confess to the crimes once caught, and who have had previous criminal convictions as well as shown other counterproductive behaviours in their work history. It is reasonable to describe these employees as 'troublemakers'. These employees are more likely to be Deliverance employees than agency or casual staff.

These offenders appear to be well known within the system and, thus, there would appear to be the opportunity for Deliverance to utilise its information more coherently in monitoring these individuals. These offences are not perpetrated by individuals with low levels of personal responsibility and commitment. In other words they are more likely to be family men with partners, though it must be pointed out that they are just as likely as the 'nothing to lose' offenders not to have children, to be quite young (under 28) and to have a short length of service.

. Vetting of Agency Staff
. How many non BP people do we have in 'F' roles + above

Nothing to lose
The offenders represented in this region are characterised by the fact that they have very little to lose should they be caught and consequently lose their jobs. They tend to be agency or casual staff, without a partner, facing some financial difficulty, and having little in the way of financial status in that they do not own a car. As with the troublemakers they tend to be without children, be under the age of 28 and have a short length of service.

These offenders are less well known in the system and are unlikely to have come to the attention of the IHIU before, by virtue of the fact that they are agency staff. Also, they are less likely to display other counterproductive behaviours or to confess when caught.

The two regions that SSA has revealed tell us that there are some individual characteristics that appear to define offenders in Deliverance. However, these are limited and it is quite apparent that many other non-offenders could also share these characteristics unless the latter are offence-specific. It was not possible to compare against a base rate in this analysis. In spite of this, these findings do suggest that there are identifiable sub-groups of employee offenders in Deliverance who are qualitatively different in their individual characteristics.

Discussion and conclusion

The first contention of this chapter was that CCD files constituted an ecologically valid source of information for analysis in this context. The reasoning behind this was that they contained information on actual offences rather than relying on the opinions of co-workers, which is the more usual method of yielding data on workplace crime. This was seen to have certain advantages (as well as disadvantages) compared to the more traditional survey approach. To summarise here, the ecological validity of CCD can be severely impaired if it lacks critical information and does not represent a full picture of the offence and its perpetrator.

Secondly, the CCD files analysed in this chapter reveal that they contain information which, if analysed appropriately, can provide valuable business intelligence. Univariate analysis (not reported here) produced a portrait of the typical Deliverance employee as age 30, male, single, graded at Deliverance level 2, with a one-in-six chance of having previous convictions, under little economic pressure, and with a mean length of service of 3.5 years. To anyone within Deliverance this would no doubt describe the vast majority of employees in their service. As in other reported findings[22] there appears to be very little to distinguish the Deliverance employee offender from his/her law-abiding workmates. Whilst this appears to offer very little help to the company in 'profiling' the typical employee offender, it does provide support for previous researchers' conclusions that employee offenders are male, in their twenties and of low tenure.

However, the SSA of some of the individual employee offender variables identified what appear to be two types of offender in action against Deliverance. Such a distinction has not previously been identified in employee offenders. These were named 'troublemakers' and 'nothing to lose offenders'. These multivariate findings are of great value. They indicate that within one organisation many different factors are having an effect at the individual level, and that these can serve to distinguish types of offenders in the workplace. They also offer support for previous researchers' hypotheses[23] that some offenders tend to have low commitment to the employing organisation and have little to lose if apprehended. Other offenders tend to make more trouble and commit other counterproductive behaviour, but these tend to be actual Deliverance employees who have more to lose if apprehended. The fact that crime against the employer appears to be the province of employees aged in their early to mid-twenties and with low length of service is supported in this data because these are common characteristics of both groups of offenders from the SSA plot. In the future there is a need to establish if there is any link between these two emergent types of offender at work in Deliverance and other factors.

The employment of CCD files in these analyses has indicated that such files have great potential for providing intelligence to Deliverance about employee offenders and for the way that Deliverance investigates them in the future. However, this situation is not perfect. Indeed, the quality of information contained within some of the CCD files can only be described as poor. There was a considerable amount of missing data, and the only standardised data consistently present was the offender's name, age, sex, and the location where he/she worked. The focus of information-gathering for the IHIU was to indicate the offender's guilt or innocence. The focus of the research, on the other hand, was to gather information about the whole domain of this activity, so it is no surprise that the two approaches are not necessarily complementary. However, the potential of this data has clearly been demonstrated in this context. To draw more solid conclusions based on a comprehensive data set Deliverance need to review their data-gathering techniques. Most of the drawbacks of employing ecologically valid data in this context, such as missing data and inconsistent recording, were present in the data set derived from the CCD files. This needs to be rectified if this data is to be of real value to Deliverance and its IHIU. However, another important part of the process is being able to see how the data can be utilised by the organisation once collected. This requires that the value of the data be demonstrated, and this chapter offers some indications of how this data can be more usefully employed. Finally, for this data to be of long-term value, the findings of this analysis really do require comparison with a non-offender sample of Deliverance employees. It may well be that the characteristics of the offenders identified here typify the majority of Deliverance employees who do not offend.

Finally, the CCD files provided this research with a sample of over 100 employee offenders. This sample involved a trawl through the IHIU files over the previous four years. This appeared to be a low level of cases for a four-year period. There may be many reasons for this lack of cases, but the apparent low frequency of apprehended employees strikes a chord of incongruence with the notion that over 30 per cent of Deliverance's losses are attributable to theft. This figure was proffered by Deliverance management at the start of the research, and there are likely to be a number of explanations for this assertion. First, this low frequency may be simply attributable to the fact that there are a few dedicated offenders stealing a high volume of parcels from their employers. However, whilst the highest estimated loss for one employee in the CCD files was £40,000–50,000, 26 per cent of the cases were for losses to the value of £250 or less. 50 per cent of the cases contained no information in this context. This hardly appears to represent a dedicated group of employees stealing high volumes of parcels, when for one business unit (out of 13), for one month, 30 per cent of loss would amount to, on average, £9000. Second, it may well be that the IHIU is not particularly adept at apprehending employee offenders. From the data in this analysis it certainly cannot be described as proactive, and this is mainly for the reason that it does not understand the importance of the data it maintains and how this can help it to reduce crime—especially if utilised in conjunction with other data sources. Complicating this matter is the fact, which was revealed through several conversations with Deliverance representatives, that IHIU offender files are not the exhaustive source of information on employee crime that they were thought to be. Indeed, a number of misdemeanours for which there were case files for some employees were not even referred to the IHIU by the depot manager in other instances. Instead, he/she preferred to deal with these as local issues and not to pass the information on to the IHIU for collation and analysis (a process which does not really occur anyway). This is a crucial source of intelligence that is not being harnessed by the IHIU, and could well have a profound effect on the outcome of some of the analyses in this chapter. Third, there is the very real possibility that there is not as much crime occurring in Deliverance as was suggested. The number of CCD files certainly did not support the level of crime suggested by management. As the research was drawing to a close it was revealed by Deliverance that an in-depth internal audit of a depot with high loss figures in the Midlands pinpointed its loss through theft (after all other factors had been taken into account) as less than four per cent. These were very revealing findings, suggesting that Deliverance management was possibly not as in touch with its inner workings as it at first appeared to be.

To conclude, this chapter has highlighted some value in the use of CCD files as a source of intelligence on employee offenders. The analysis of the files has revealed support for the findings of other researchers, that employee offenders tend to be young, of low tenure and with little responsibility. In particular, analysis of these files revealed two types of offender to be in operation against Deliverance: 'troublemakers' and 'nothing to lose offenders'. The identification of these offender types has specific implications for Deliverance in that their short-tenured, temporary, younger staff appear to be particularly vulnerable to the temptation to steal. At the same time elements of their tenured staff also offend against the organisation, but these offenders more commonly have other counterproductive behaviours in their work repertoire. This has specific implications for Deliverance in that treating these two types of offender in the same way may prove counterproductive in the long term. These findings confirm the arguments of previous researchers that employee offenders have less commitment to the organisation because they have less to lose if apprehended, though in this context the findings have been drawn from an actual employee sample rather than a survey of a workforce. However, this use of an actual employee sample will only be an effective source of intelligence to the organisation when management understands its potential value, prepares more comprehensive files of information and compares the findings of this research with a non-offending sample.

Notes

1 Nick Dodd is currently a Consulting Director, Behavioural Science & Change Solutions Ltd, Ipswich, Suffolk; email: nickdodd@bscsolutions.co.uk.

2 Horning, D.M. (1970) Blue Collar Theft: Conceptions of Property, Attitudes toward Pilfering, and Work Group Norms in a Modern Industrial Plant. In Smigel, E.O. and Ross, H.L. (eds) *Crimes Against Bureaucracy*. London: Van Nostrand Reinhold; Hollinger, R.C. and Clark, J.P. (1982) Employee Deviance: A Response to the Perceived Quality of the Work Experience. *Work and Occupations*. Vol. 9, pp 97–114.

3 Mars, G. (1994) *Cheats at Work*. Aldershot: Dartmouth; Cressey, D.R. (1953) *Other People's Money*. Glencoe, IL: Free Press.

4 Dodd, N.J. (1998) Applying Psychology to the Reduction of Insurance Claim Fraud. *Insurance Trends*. No. 18. London: Association of British Insurers; Robertson, A. (2000) Theft at Work. In Canter, D. and Alison, L. (eds) *Profiling Property Crimes*. Aldershot: Ashgate.

5 Mars, G. (1974) Dock Pilferage. In Rock, P. and McIntosh, M. (eds) *Deviance and Social Control*. London: Tavistock; Ditton, J. (1977) *Part-time Crime: An Ethnography of Fiddling and Pilferage*. London: Macmillan.

6 Felson, M. and Clarke, R.V. (1997) Introduction: Business and Crime. In Felson, M. and Clarke, R.V. (eds) *Business and Crime Prevention*. New York: Criminal Justice Press; Burrows, J. (1997) Criminology and Business Crime: Building the Bridge. In Felson and Clarke, ibid.

7 Canter, D.V. (1994) *Criminal Shadows*. Aldershot: Dartmouth; Hollin, C. (1996) *Psychology and Crime*. London: Routledge.

8 Jones, D.C. (1972) Employee Theft in Organizations. *Society for the Advancement of Management*. Vol. 37, pp 59–63; Moretti, D.M. (1986) The Prediction of Employee Counterproductivity through Attitude Assessment. *Journal of Business and Psychology*. Vol. 1, No. 2, pp 134–47; Terris, W. (1985) Attitudinal Correlates of Employee Integrity. *Journal of Police and Criminal Psychology*. Vol. 1, No. 1, pp 60–88.

9 See Franklin, A.P. (1975) *Internal Theft in a Retail Organization: A Case Study*. Ann Arbor, MI: University Microfilms; Hollinger, R.C. and Clark, J.P. (1983) *Theft by Employees*. Lexington, MA: Heath; Hollinger, R.C., Slora, K.B. and Terris, W. (1992) Deviance in the Fast-food Restaurant: Correlates of Employee Theft, Altruism and Counterproductivity. *Deviant Behavior: An Interdisciplinary Journal*. Vol. 13, No. 2, pp 155–84; Robin, G.D. (1969) Employees as Offenders. *Journal of Crime and Research on Crime and Delinquency*. Vol. 6, pp 17–33; Robertson, A.R. (1993) *A Psychological Perspective on Blue-Collar Workplace Crime*. Unpublished MSc dissertation, University of Surrey.

10 See Dodd, N.J. (1998) Applying Psychology to the Reduction of Insurance Claim Fraud. *Insurance Trends*. No. 18. London: Association of British Insurers; Cressey, op cit.

11 See Greenberg, J. (1990) Employee Theft as a Reaction to Underpayment Inequity: The Hidden Cost of Pay Cuts. *Journal of Applied Psychology*. Vol. 75, No. 5, pp 561–68; Hollinger and Clark (1983) op cit; Laird, D.A. (1950) Psychology and the Crooked Employee. *Management Review*. Vol. 39, pp 210–15; Szwajkowski, E. (1989) Lessons for the Consultant from Research on Employee Misconduct. *Consultation*. Vol. 8, No. 3, pp 181–90; Tucker, J. (1989) Employee Theft as Social Control. *Deviant Behavior: An Interdisciplinary Journal*. Vol. 10, No. 2, pp 319–34.

12 See Cressey, op cit; Harrell, A.W. and Hartnagel, T. (1976) The Impact of Machiavellianism and the Trustfulness of the Victim on Laboratory Theft. *Sociometry*. Vol. 39, No. 2, pp 157–65; Hollinger and Clark (1983), op cit; Paul, F.M. (1982) Power, Leadership and Trust: Implications for Counselors in Terms of Organizational Change. *Personnel and Guidance Journal*. Vol. 60, No. 9, pp 538–41.

13 Ditton, op cit; Hollinger, R.C. (1986) Acts against the Workplace: Social Bonding and Employee Deviance. *Deviant Behavior: An Interdisciplinary Journal*. Vol. 7, No. 1, pp 53–75; Hollinger and Clark (1983) op cit; Mars, op cit; Murphy, K.R. (1993) *Honesty in the Workplace*. Pacific Grove, CA: Brooks/Cole.

14 Murphy, op cit; Paajanen, G.E. (1988) *The Prediction of Counterproductive Behavior by Individual and Organizational Variables*. Unpublished PhD thesis, University of Minnesota.

15 Hollinger and Clark (1983) op cit.

16 See Sackett, P.R. (1985) Honesty Research and the Person-situation Debate. In Terris, W. (ed.) *Employee Theft*. Chicago, IL: London House; Sackett, P.R. and Harris, M.M. (1985) Honesty Testing for Personnel Selection: A Review and Critique. In Bernardin, H.J. and Bownas, D.A. (eds) *Personality Assessment in Organizations*. New York: Praeger.

17 Krippendorf, K. (1980) *Content Analysis.* Beverly Hills, CA: Sage.

18 Wilson. M. (1998) Structuring Qualitative Data: Multiple Scalogram Analysis. In Breakwell, G.M., Hammond, S. and Fife-Shaw, C. (eds) *Research Methods in Psychology.* London: Sage.

19 Canter, D.V. (1985) *Facet Theory: Approaches to Social Research.* New York: Springer Verlag; Shye, S., Elizur, D. and Hoffman, M. (1994) *Introduction to Facet Theory.* London: Sage.

21 Lingoes, J.C. (1974) *The Guttman-Lingoes Non-metric Program Series.* Unpublished MA thesis, University of Michigan.

22 Ditton, op cit; Dodd, op cit; Mars, op cit.

23 Hollinger *et al*, op cit; Bamfield, J. (1998) A Breach of Trust: Employee Collusion and Theft from Major Retailers. In Gill, M. (ed.) *Increasing the Risk for Offenders.* Vol. II, *Crime at Work.* Leicester: Perpetuity Press.

Chapter 12

Tackling Shrinkage in the Fast Moving Consumer Goods Supply Chain: Developing a Methodology

Adrian Beck, Charlotte Bilby and Paul Chapman[1]

Shrinkage for retailers and suppliers of fast moving consumer goods (FMCG) continues to be a significant problem, and one that seems resilient to remedial action. In an effort to respond to the apparent failure of existing approaches to loss prevention, this article introduces a 'process-orientated' approach to tackling shrinkage. The approach consists of a prescriptive series of seven steps that form a stock loss reduction 'roadmap', together with associated techniques and tools for undertaking each step.

Introduction

Recent research has once again demonstrated the extent of the problem of shrinkage for retailers and suppliers of fast moving consumer goods (FMCG).[2] Losses in this €824.4 billion[3] industry were found to equate to 2.31 per cent of turnover, 1.75 per cent for retailers and 0.56 per cent for manufacturers, resulting in an annual bill for shrinkage of €18 billion.

In some respects, there is nothing new about attempting to quantify the cost of stock loss.[4] However, it is increasingly clear not only that the problem is enormous, but that the issue seems resilient to remedial action. This failure calls into question the effectiveness of the world-wide market for security 'solutions', for example closed circuit television (CCTV), electronic article surveillance (EAS), guarding and store detectives, and the effectiveness of the multi-billion-euro global security industry. It could be argued that this expenditure 'keeps the lid' on losses and contains them to relatively acceptable levels. However, this approach seems only to be an argument of convenience—although proving this to be the case is currently beyond most established methodologies.

Recent research has shown that only 20–40 per cent of loss is accounted for by external theft.[5] Hence many of the technology-led solutions, such as EAS, are only capable of impacting on a minority of modes of stock loss. They make little or no impact on delivery and warehouse staff theft, or on losses caused by products going out of date due to poor company ordering procedures. Accordingly, this chapter seeks to provide a new methodology that can enable shrinkage practitioners to begin to adopt a more studied approach in dealing with the problem. It is based upon original research carried out throughout the FMCG sector in Europe by academics specialising in supply chain management and security management.

This approach, termed here a 'roadmap', charts the various steps that need to be taken in order for a more holistic and systematic approach to be taken. This roadmap aims to supersede existing approaches, which characteristically involve partial, piecemeal and for the most part poorly conceived practices. In doing so, the intention is to lead organisations to face up to their large and growing bill for shrinkage, while also presenting them with a rich opportunity for dramatic improvements in profitability.

Methodology

This study used a methodology based upon two approaches: a review of the existing literature, and investigative fieldwork exploring 12 FMCG supply chains. The analysis of these supply chains required an investigation of 17 companies, including manufacturers, distributors and retailers, on 41 sites in 10 European countries. To follow the flow of products through FMCG supply chains in a structured manner required both a systematic and a systemic approach. Being systematic, this research was methodically arranged and undertaken according to a plan. The systemic nature of the work is seen in the focus on each supply chain as a single process, by integrating research undertaken at the separate sites along each supply chain.

Each site visit, whether to a manufacturing company, distributor or retailer, consisted of structured interviews with members of the senior management team and with the managers responsible for security, logistics and stock loss. In addition to the interviews, goods were physically followed through sites in order to document the supply chain process and to identify the practices and procedures used to facilitate this process and to control losses. The investigation made use of 'failure modes and effect analysis' (FMEA) to examine the various ways that a process may fail, and to determine the effect of the different failure modes. Through focusing attention on stock loss, and by using the FMEA technique, it was possible to understand the following:

- the ways in which the supply chain process can fail and allow stock to be lost;
- the severity of the loss should such a failure occur;

- the likelihood of it occurring; and
- the ability to detect that it has occurred.

The data collated from the process analysis and the interviews allowed a composite model of the supply chains to be constructed. This provided the basis for the analysis of the supply chains' performance and for drawing conclusions.

The roadmap

In theory, the concept of stock loss reduction is simple. It can be described in terms of the following three steps:

- making stock highly visible so that loss is immediately noticed;
- quickly identifying the cause of the loss; and
- implementing preventative solutions to resolve the cause of the loss and thus to prevent a reoccurrence of the problem.

Difficulties in implementation arise for a wide range of reasons. Not least are the complexity of the sector, the absence of reliable data on the extent and nature of the problem, and a lack of co-operation, both within and between companies in the supply chain, in developing shared solutions. Underpinning these issues are the difficulties associated with managing change, particularly the lack of a clear method for undertaking stock loss reduction projects.

Stock loss reduction principles
In order to overcome the inhibitors to stock loss reduction, the roadmap is constructed around the application of three key principles—it is:

- holistic;
- systematic; and
- collaborative.

These three principles underpin the structure of the roadmap, and also the method for applying it. The roles of the three principles are described below.

Holistic
Resolving stock loss requires businesses to take a holistic approach to identifying and analysing stock loss problems from a supply chain perspective. Supply chains consist of processes made up of diverse, interlinked tasks and hence analysing the

processes opens up the 'black box' of the supply chain. Taking this process-orientated approach to the investigation of both individual activities and whole organisations allows both the wood and the trees to be analysed. It also ensures that the realities of work practices are considered in parallel with the supply chain's overall function.

Systematic

The complexity inherent in supply chains requires their investigation to be undertaken in a systematic manner. This requires planning the investigation methodically and undertaking the work according to a plan. For example, where a site visit is required, whether to a manufacturing company, distributor or retailer, the visit needs to consist of structured data-gathering, through interviews, to collate the following information and data:

- stock loss statistics;
- the company's methods of identifying stock loss;
- the extent of staff and external theft;
- the company's methods of reducing stock loss;
- the justification for these methods; and
- the impact of loss prevention methods on sales.

In addition, an understanding of the structure of material-handling and information exchange processes is gained through physically following goods through the material-handling processes. This is undertaken in order to document the activities and associated practices and procedures used to facilitate this process and to control loss.

Collaborative

Collaboration within a company provides the means of resolving problems which are beyond the scope of individual functions. Likewise, collaboration between companies resolves communal problems beyond the scope of the separate companies. The results of this work will be to:

- design loss prevention solutions into the fabric of processes and facilities;
- ensure that these solutions effectively contribute to total supply chain efficiency, instead of solving a problem at one end of the chain only; and
- implement simpler and cheaper controls to minimise loss.

Structure of the stock loss reduction roadmap
The roadmap seeks to overcome the inhibitors to successful loss prevention by employing a stock loss reduction method consisting of both:

- a general approach to stock loss reduction; and
- specific tools and techniques for dealing with specific problems.

The roadmap is intended as a manual or guide, describing the overall activities that need to be undertaken in order to reduce stock loss. This guide consists of a general approach made up of the steps a company needs to follow, together with techniques and tools to help undertake each phase and to deal with problems that may be encountered. The general approach that forms the heart of the guide is shown in Figure 1. The structure is systematic, and provides a means for planning and undertaking stock loss reduction projects while guiding users towards continuous improvement through the cycle.

Figure 1. The shrinkage roadmap

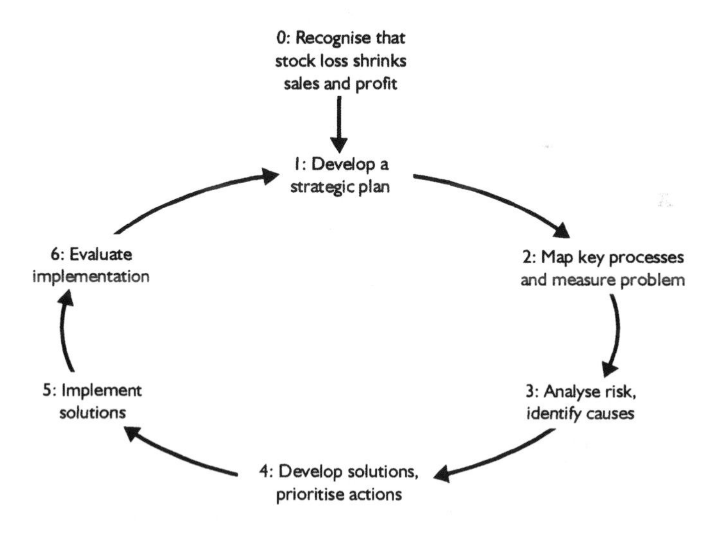

The roadmap is designed to enable both problem diagnosis and solution implementation. The structure consists of a sequence of steps whereby, through well-planned investigation, pressing business needs are identified. These needs are responded to by undertaking small-scale experiments that explore the stock loss problem, identify likely causes and develop appropriate solutions. The solutions are tested in trials where their effectiveness is assessed. Where a solution is found to be successful, it is then implemented widely and the business's practices standardised around it.

This approach to stock loss reduction recognises, given the uniqueness of each business environment where stock loss occurs, that the prescription of a single, 'right' strategy for reducing stock loss is inappropriate. Instead, whilst a basic structure is provided, the approach is intended to be tailored to match prevailing circumstances in order for it to be effective.

The approach described in this chapter provides a means for involving all company employees in stock loss reduction. Widespread participation provides knowledge of current practices, such as stock control, and helps build commitment that allows changes to be implemented.

To help undertake the steps of the general approach, a number of techniques and tools are recommended. These have been chosen to aid communication and understanding. This list is not comprehensive, and prospective users should introduce their own tools where they find them to have a better fit in the prevailing local circumstances.

Figure 1 shows the roadmap as consisting of six sequential steps, that join to form a circle of diagnosis and implementation. These steps are described below in more detail, along with a number of techniques and tools that can be used to help accomplish the objectives of each step and overcome problems that may be encountered.

Step 0: recognise that stock loss shrinks profit and sales

As related research has shown, the FMCG sector in Europe suffers significant losses each year through shrinkage (€18 billion).[6] These losses occur all along the supply chain: from point of manufacture, throughout the distribution process, to the point of sale. While €10 billion of this loss cannot be accounted for, the effects of shrinkage are clear: the shopper suffers through added cost and poorer service. The outlook for retailers, manufacturers and consumers concerning shrinkage is bleak unless action is taken quickly.

Whilst current levels of stock loss are significant, future levels are likely to increase. Food retailers continue to expand the number of non-food items they carry, such as CDs, clothes and electrical items, which are products perceived by them to be at greatest risk of theft. In general, current attempts to address shrinkage are characterised by a heavy reliance upon reactive strategies that are only triggered when a particular problem becomes intolerable. Such knee-jerk and insular reactions not only fail to resolve the causes of loss but can also have a detrimental effect on the profitability of a company. For instance, adopting 'defensive merchandising' can cause problems with replenishment and availability. Hence the first step in the roadmap is two-fold:

- wake up to the magnitude of the stock loss problem; and
- recognise that current approaches to stock loss reduction generally do not work.

Step 1: develop a strategic plan

Current efforts to contain stock loss are at best piecemeal, with few companies operating an organisation-wide approach to resolving the problem. Companies need to recognise that traditional approaches, that are only tasking security, audit or health and safety departments with the challenge, are not effective in tackling the structural deficiencies in business operations that are the root cause of stock loss. Instead, companies need to change their approach to resolving stock loss and make use of a wider range of skills and resources. In changing their approach, they need to choose one that is systematic, holistic and collaborative.

Applying the principle of being 'holistic'

Effective stock loss reduction requires companies to be holistic by identifying, for the supply chain as a whole, where problems occur and can best be resolved. Such work requires collaboration along supply chains between suppliers, distributors and retailers as well as across the FMCG sector as a whole. Only when internal and external problems are considered together can comprehensive analysis be undertaken to deliver early, tangible results. Currently, companies are simply not taking advantage of the opportunities to share expertise with either their competitors or suppliers, or indeed internally.[7] For example, only one-half of all retailers and manufacturers are working together to tackle stock loss. This clearly demonstrates that problems affecting the whole of the supply chain are not being addressed in a holistic manner.

Applying the principle of being 'systematic'

In general, stock loss is not currently approached in a systematic manner. In order to break away from the culture of half-truths and anecdotes, a systematic approach provides the way for a company to quantify and prioritise its problems, to analyse the causes of these problems and to direct its available resources to the most cost-effective solutions. Finally, the true effectiveness of these solutions needs to be determined after their implementation and this information then used to guide future investment. Hence the first step in the systematic approach to stock loss reduction is planning, and the last step is the evaluation of the effectiveness of actions.

Planning is based on clear, realistic, attainable objectives, accompanied by criteria for knowing when these objectives are met. This requires the project team responsible for delivering loss reductions to have answers to the following questions:

- What is the supply chain process which is to be improved?
- When does this process start and finish?
- What are the goals of the stock loss reduction activity?

- When is the date by which some benefits must be felt?
- What are the attributes of the ideal supply process?
- What are the constraints on improvement?
- What are the stock loss threats faced by the company?

The answers to these questions guide the project team's activities towards achieving their goals. Starting the project in this way is especially important in cross-functional/inter-company projects, where the effectiveness and efficiency with which project resources are used dramatically improves with up-front investment in planning.

Step 2: map key processes and measure the problem
Reducing stock loss begins with a rigorous diagnosis of the problem. This diagnosis starts by understanding the nature of the losses, and then identifies their causes. Understanding the current operational system and processes is also the first step in gaining widespread recognition of the problem and in establishing the need for change within an organisation. The act of creating a business process model that identifies the source of stock loss can develop the critical momentum required to change existing behaviour.

A process-led approach to reducing stock loss applies process analysis to the stock loss problem, and emphasises prevention. This provides the means to identify and prevent system and procedural losses, which also reduces loss from theft by removing the opportunity to abuse deficient systems.

The starting point for taking a process-orientated approach to diagnosing stock loss is to undertake process mapping and measurement. These two techniques are described below.

Process mapping. This is a technique used to detail business processes that focus on the important elements influencing behaviour, allowing the business to be viewed at a glance.[8] Mapping and measuring a process establishes the performance baseline that enables the effectiveness of solutions to be measured. An example of a top-level supply chain is depicted in Figure 2. This shows two supply chains from Gillette in the UK to two of their customers, ICA in Sweden and Tesco in Hungary.

Whilst this gives an understanding of the total supply chain, it provides few details. In order to show more in-depth information, a more rigorous process map is required. Simple flow-charting techniques are often the most appropriate technique to use when process mapping for the first time.

Figure 2. Gillette supply chains to ICA, Sweden and Tesco, Hungary

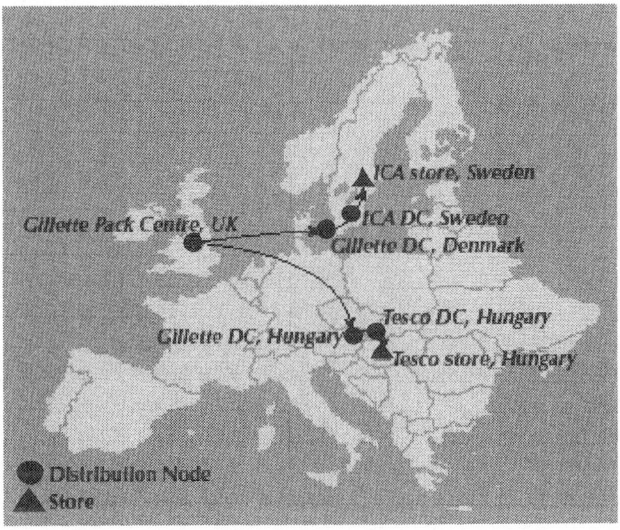

The data for creating a process map is best collated by physically following products as they pass along the supply chain. This involves visiting each site that the products pass through and documenting the steps involved in receiving, storing and dispatching them. Figure 3 is an example of a process map showing the flow of products through a distribution centre.

Figure 3. Distribution centre product flow chart

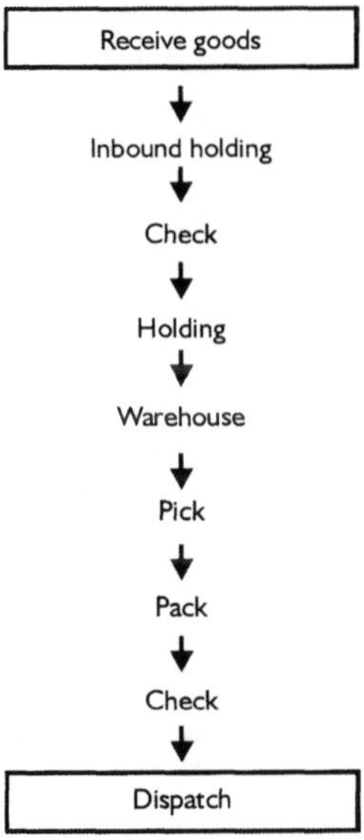

Receive goods
↓
Inbound holding
↓
Check
↓
Holding
↓
Warehouse
↓
Pick
↓
Pack
↓
Check
↓
Dispatch

This presents the steps involved in receiving, storing and dispatching goods in a simple graphical format. Even when the flow chart does not provide a complete or totally accurate model of a process, it will significantly add to understanding by promoting and communicating a process-orientated approach to improvements.[9]

Hot products. Processes and systems usually contain a large number of separate products. Rather than map all the various routes taken by all the different items it is appropriate in the initial cycle of analysis to focus on 'hot' products.[10] These products illustrate general features of the supply chain and expose major problems inherent within it. The hot product is a concept that many retailers and manufacturers are familiar with, and generally refers to those products most attractive to thieves. If retailers and manufacturers were to gain a better idea of what makes a product hot, then this could, by reducing the levels of the theft element within stock loss, help dramatically reduce levels of shrinkage within the whole supply chain.

Current approaches to categorising hot products are usually based on the perceptions of the security department, as methods are not always in place to analyse the true nature of stock loss. As the recent ECR Europe data shows, retailers can identify only 41 per cent of their losses, and manufacturers are aware of only 59 per cent of the losses they suffer.[11] However, the study also found that non-food products are perceived to be most at risk of theft, particularly tobacco goods, videos, CDs, DVDs, beers, wines and spirits, health and beauty products, and electrical goods. These items increasingly feature within grocery stores, so the need to control their loss may be the trigger for retailers and their suppliers to work collaboratively to deal with this problem. While the concept of hot products refers mainly to items that are stolen, lessons learnt from closely monitoring their progress throughout the entire supply chain may have more generalised benefits for improving the processes used to move these and all other products.

Measuring the problem. The ECR Europe shrinkage study also found that currently within the FMCG sector most retailers keep records of supplier fraud and process failures, but few keep computerised records of internal and external theft, at either a company or a store level. Manufacturers were found to keep computerised records of process failures, but few recorded any form of theft on a computerised system. The majority kept either no record whatever or only paper files.[12]

It is only through the use of computerised databases that trends can be identified and a more information-led, systematic approach adopted to deal with all the elements that account for shrinkage. There are a number of fundamental measures of stock loss that are required in order to determine the What? How? and When? for each stock loss incident. From a supply chain perspective, the following basic measures need to be collected:

- the level of loss in deliveries to a site;
- the level of loss at a site; and
- the level of stock loss in deliveries from a site.

These measures allow a 'top-level' assessment of the extent and location of stock loss across a supply chain. In addition, it is necessary to collect data on the following factors:

- the type of incident;
- how the incident occurred; and
- when it occurred.

This data allows the nature of the type of stock losses suffered to be better understood. The points in the supply chain where these measures should be taken are shown in Figure 4.

Figure 4. Points of measurement across the FMCG supply chain

(1) The level of loss in deliveries to a site (2) The level of loss from a site
(3) The level of stock loss in deliveries from a site

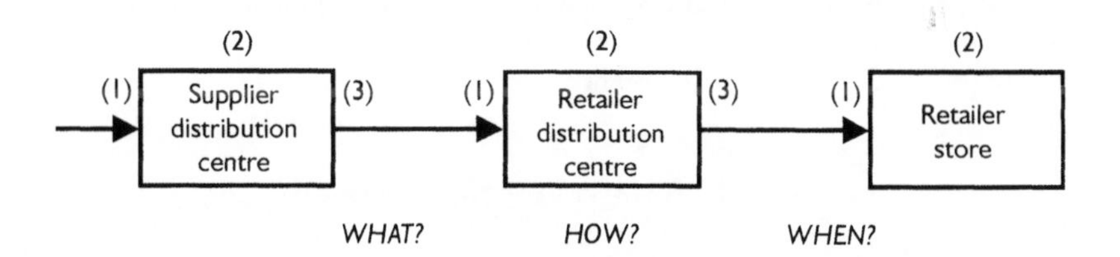

In addition to where the loss has taken place, companies need to develop systems that track how and when particular incidents occur; and they need to do this for all types of stock loss. They need to develop an approach that is systematic, and part of an on-going process for collecting and collating timely and useful information that describes patterns, trends and information on stock loss throughout the supply chain—from first delivery to final check out.

Step 3: analyse risk, identify causes and prioritise actions

Having mapped and measured the current operation, this data should be analysed to understand and describe exactly what is wrong. 'Failure modes and effect analysis' (FMEA) is a tool for assessing the various ways that a process may fail and to determine the effect of the different failure modes.[13] Through focusing attention on stock loss, and by using the FMEA technique, it is possible to understand the following:

- the ways in which a process can fail and allow stock to be lost;
- the severity of the loss should a failure occur;
- the likelihood of a failure occurring; and
- the ability to detect that a failure has occurred.

Collation of this information quantifies the risks of stock loss for the supply process as a whole. This quantification allows the risks to be ranked in order of severity, and hence the key risks are identified through a systematic and robust method.

Cause and effect analysis. This can be applied in two ways:

- by reactively analysing particular incidents of stock loss; and
- by proactively analysing the key risks of stock loss quantified in the FMEA exercise.

Cause and effect analysis benefits from a long and successful history of application in the investigation of quality problems, and is fairly simple to understand and use.[14] Having identified specific symptoms of poor performance, the cause and effect diagram, shown below in Figure 5, is an effective way of capturing possible contributing causes to that performance.

Figure 5. A cause and effect diagram

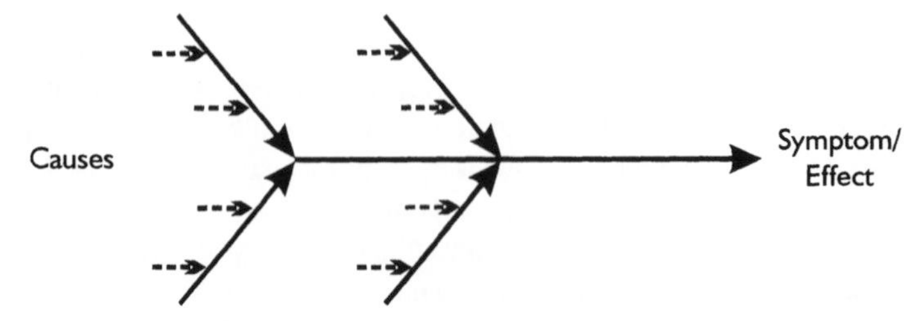

This diagram is especially useful when employed to structure the outcome of a brainstorming session with the project team, when they contribute their findings, experience and understanding.[15] The main spines of the diagram are given broad headings around which causes for the symptom of a problem are grouped; however, the choice of these headings is fairly arbitrary.

To focus effort, the major causes of problems need to be identified from amongst the trivial many. This should be achieved quantitatively through the collection of data, using check sheets to determine the number of incidents associated with each of the causes that have been suggested. However, it is possible to get the project group members to identify many of the most significant problems from their experience. In Figure 6, three causes of stock loss have been highlighted as being the most significant ones for this particular site. These are the causes that will be investigated further.

This approach follows the Pareto Principle that the 'vital few' causes are responsible for the bulk of problems.[16]

Figure 6. Stock loss cause and effect diagram with three significant causes highlighted

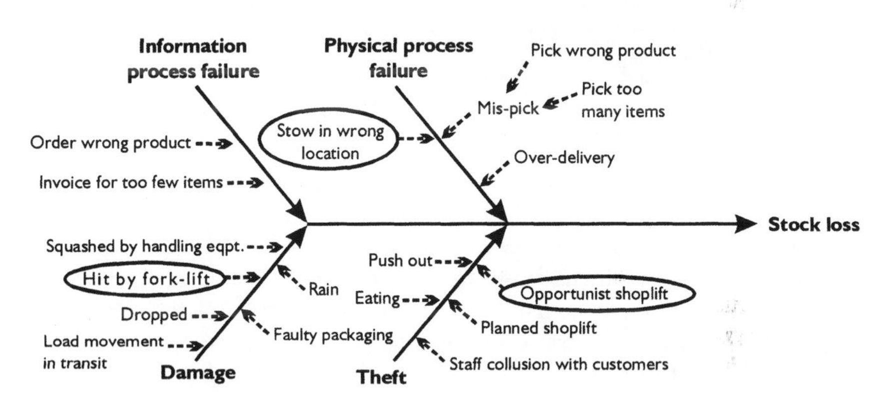

The five whys. Initial ideas about the causes of problems are unlikely to identify underlying root causes. Having tidied the initial ideas and focused on the significant ones, deeper cause-and-effect structures need to be identified. A technique to explore underlying causes of problems, not those that are first perceived, is the 'five whys' technique.[17] This explores the underlying causes of losses as fully as possible by repeating the question 'why?'[18] The technique is illustrated overleaf in Figure 7.

The five-whys technique also illustrates that the causes of loss are likely to be many and varied. In order to focus effort onto the key underlying causes, the Pareto Principle should be applied once more, and the vital few causes of loss highlighted from amongst the trivial many.

Identification and understanding of root causes concludes the diagnosis of the causes of stock loss, and starts the 'remedial journey' where solutions to these problems are sought.[19]

Figure 7. A 'five whys' diagram used to investigate the root causes of effects

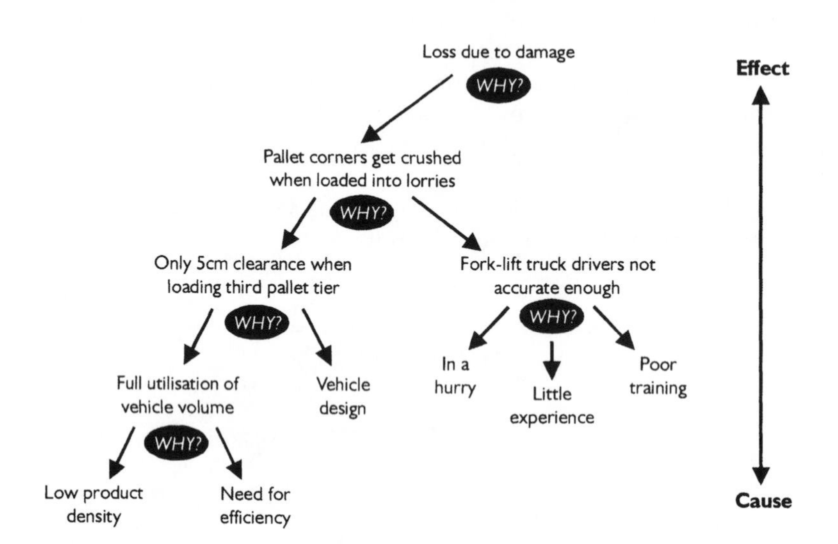

Step 4: develop solutions and prioritise actions
Organisations traditionally start their stock loss reduction efforts at this, the solution stage. It is not uncommon to find a great solution and then search for a problem to apply it to. The systematic process of investigation described in Steps 0–3 of the roadmap requires stock loss problems to be first investigated, and their causes identified, before the development of solutions that resolve these causes, and hence reduce loss. These solutions are usually extremely context-specific, and it is therefore not possible simply to prescribe solutions without considering the findings of the investigative work. However, it is important to note that having identified a set of appropriate solutions, these can be implemented through one of two general approaches: 'clean sheet', or the renovation of existing operations.

The clean sheet approach sets existing systems to one side and starts afresh. It recognises that current practices are beyond salvage, and have no further use. Renovating existing processes builds upon the capabilities that have underpinned the historical success of the organisation. This latter approach requires these capabilities to have retained some value, which may not be the case.

New processes and systems should be accompanied by newly designed performance measurement systems. In the same way that processes are redesigned to deliver their objectives, so the performance measurement systems also need to be redesigned to monitor and control the new processes. Such a system requires a suite of measures. These must reflect the range of factors important to the organisation that the improvement project needs to enhance. Considering them in harmony, for instance by using a 'balanced scorecard',[20] promotes improvements across a broad front, or at least ensures that performance is maintained for the basket of measures whilst driving progress in one key measure. Where performance levels essential to future success have been identified, but the process design that delivers them is not understood, benchmarking can be a useful technique to help overcome this.[21] A benchmarking exercise helps identify the processes used in other organisations that enable them to achieve superior performance, for example by benchmarking against other hot products or items from other product categories.

Step 5: implement solutions

In a similar manner to the approach used to plan the project investigation, the implementation of the solution that will reduce stock loss requires project planning. Successful projects require a sponsor to be responsible for delivering the benefits of the project. To achieve success the sponsor, usually a senior manager, needs to ensure that the project team constructs a clear and robust business case. This business case defines what is to be delivered, the benefits this will bring and the resources required.

The creation of a project plan is necessary to target resources to achieve the desired objectives within time and cost limitations. At a top level, a project plan consists of a sequence of steps that need to be undertaken. A typical set of steps is described below:

- identify the overview tasks needed to complete the project;
- show the interrelationships between tasks, and the sequence in which they can be undertaken, on a network diagram;
- estimate the types and amount of effort needed to complete these tasks;
- calculate the resource profile over time to complete the project;
- identify potential risks to successful project delivery;
- mitigate risks or plan contingency;
- iterate the plan to match it against resource availability;
- secure resources; and
- put in place procedures for evaluation.

Evaluating the effectiveness of the stock loss reduction cycle provides information that guides the direction of the next cycle. Stock loss reduction needs to be ongoing to ensure loss reduction efforts are compatible with developments across the supply chain, and to counter the resourcefulness of criminals.

Step 6: evaluate the implementation
The stock loss reduction project ends with an effective solution in place. However, this is not the end of stock loss reduction as a whole. From the organisation's perspective, evaluation of one project is important in order to determine the success of the solution and to guide future projects. The review is therefore the final step of one project and the first step of the next.

The ability to sustain significant improvements in stock loss over long periods of time rests on the capability to learn from experience, and to ensure that the company accesses the wide range of developing tools at its disposal. Therefore this review of the implementation must be objective. All too often reviews are undertaken with the aim of justifying the work that has been done, and fail to provide an honest appraisal of what solution worked and why. Accordingly, the evaluation should be rigorous, robust and led by somebody who can take an objective view, independent of equipment providers and those who may have commissioned the project in the first instance. He/she needs a clear mandate to assess the performance of the implemented solution and compare this against the level of performance originally planned. This assessment should consider how the implementation of solutions was justified, for example by the use of a cost/benefit analysis. This information provides the feedback that allows the stock loss reduction team to consider objectively the effectiveness of the approach the project team took to reducing stock loss and of the specific solutions they implemented.

In addition, the aim of this feedback phase is to identify whether any further action is required before the current project can be signed off, and to gain a better appreciation of successful approaches and solutions that might be applied during future projects. It should be noted, however, that the evaluation process may need to be ongoing, as the performance of an initiative can change as its 'environment' alters. For instance, criminals may gradually find ways of defeating the newly adopted approach, or changes in product range or levels of staffing might reduce its effectiveness. Therefore, periodic reviews of newly adopted measures may need to be carried out in order to gauge their effectiveness over time and to evaluate whether any corrective measures need to be taken.

Conclusion

The data from a plethora of studies has shown that the problem of shrinkage is massive and growing, accounting for a significant percentage of the annual turnover of companies throughout the FMCG industry. It also shows that to date most efforts to deal with the problem have been piecemeal and partial, relying upon hearsay and guesswork.

The stock loss reduction roadmap is an attempt to begin imposing a greater degree of rigour on the stock loss process, through a more holistic and systematic methodology. It is not intended to be a prescriptive solution to the problem of shrinkage, but rather provides a means of organising stock loss reduction efforts; it is aimed at helping practitioners to think through and develop a more strategic approach. Where it achieves this aim it seems clear that the shrinkage reduction industry will have taken a dramatic step forward.

Notes

1 Adrian Beck is a Lecturer at the Scarman Centre, University of Leicester; email: bna@le.ac.uk; Charlotte Bilby is a Lecturer at the Centre for Applied Psychology, University of Leicester; email: calb1@le.ac.uk; Paul Chapman is a Senior Research Fellow at the Cranfield School of Management; email: paul.chapman@cranfield.ac.uk.

2 The term 'fast moving consumer goods sector' is used here to mean those retailers and their suppliers who provide a range of goods sold primarily through supermarkets and hypermarkets. The core of their business is providing 'essentials' such as various fresh and processed foodstuffs, but they also stock a wide selection of other goods, including health and beauty products, tobacco, alcohol, clothing, some electrical items, baby products and more general household items. Examples of FMCG retailers include Auchen, Carrefour, Coop Italia, ICA, Interspar, Tesco and Walmart. Examples of FMCG manufacturers include Allied Domecq, Gillette, Johnson and Johnson, Proctor and Gamble, and Unilever. In the USA, this sector is also referred to as the 'consumer packaged goods sector'.

3 M+M Euro Trade (2000) *Trade Structures and the Top Retailers in the European Food Business*. Frankfurt: M+M Euro Trade.

4 The terms 'shrinkage' and 'stock loss' will be used interchangeably throughout this article.

5 Beck, A., Bilby, C. and Chapman, P. (2002) Shrinkage in Europe: Stock Loss in the Fast-moving Consumer Goods Sector. *Security Journal*. Vol. 15, No. 4, pp 25–39.

6 Beck A. and Bilby, C. (2001) *Shrinkage in Europe: A Survey of Stock Loss in the Fast Moving Consumer Goods Sector*. Brussels: ECR Europe.

7 Ibid.

8 Soliman, F. (1998) Optimum Level of Process Mapping and Least Cost Business Process Re-engineering. *International Journal of Operations and Production Management*. Vol. 18, Nos. 9/10, pp 810–6.

9 Maull, R.S., Childe, S.J., Bennett, J., Weaver, A.M. and Smart, A. (1995) *Report on Process Analysis Techniques*. EPSRC Working Paper WP/GR/J95010-3. Swindon: Engineering and Physical Sciences Research Council.

10 See Clarke, R.V. (1999) *Hot Products: Understanding, Anticipating and Reducing Demand for Stolen Goods*. Home Office Police Research Series, No. 112. London: HMSO.

11 Beck *et al*, op cit.

12 Ibid.

13 Stamatis, D.H. (1995) *Failure Mode and Effect Analysis*. Milwaukee, WI: ASQC Quality Press.

14 Juran, J.M. and Gryna, F.M. (1988) *Juran's Quality Control Handbook.* 4th edn. New York: McGraw-Hill.

15 Ishikawa, K. (1990) *Introduction to Quality Control*. London: Chapman and Hall.

16 Juran, J.M. and Gryna, F.M. (1993) *Quality Planning and Analysis*. 3rd edn. New York: McGraw-Hill.

17 Bicheno, J. (1998) *The Quality 60*. Buckingham: PICSIE Books.

18 Ishikawa, op cit.

19 Juran, J.M. (1989) *Juran on Leadership for Quality*. New York: Free Press.

20 Kaplan, R.S. (1996) *The Balanced Scorecard: Translating Strategy into Action.* Boston, MA: Harvard Business School Press.

21 Camp, R.C. (1989) *Benchmarking: The Search for Industry Best Practices that Lead to Superior Performance*. New York: Quality Press.

Chapter 13

Managing Terrorist Targeting of Financial Centres: The IRA's City of London Campaign

John Gearson[1]

In the campaigns waged by the Provisional Irish Republican Army (hereafter 'the IRA') against British rule in the province of Northern Ireland, attacking the mainland, and London in particular, was always seen as a useful way of bringing home to the British people and their political leaders the costs of remaining in Ireland. In what turned out to be the organisation's final mainland bombing campaign to date, the IRA stumbled on a new and vulnerable flank in the British state—the financial services sector's vulnerability to dislocation by large vehicle-borne explosive devices, the potentially enormous insurance costs associated with the bombing of expensive office space in financial districts, and the symbolic effect of doing so. In two major bombings in the City of London in the early 1990s—the largest ever seen on the mainland in peacetime—the IRA presented the City Corporation with an unprecedented challenge: to secure the City from terrorist attack as best it could, while reassuring home and overseas investors and existing occupants of office space in the district that the City had not become an impossible place in which to do business. The City of London embarked upon a series of physical security measures and public reassurance on a scale never before seen in Britain, while the IRA attempted to breach the developing security regime and achieve another 'spectacular' against the British state, and against the financial services sector in particular. The City, while not the most often targeted part of London, or the mainland, became the battleground for the final stages of the IRA's military campaign in Britain. The lessons learned by the City of London were grimly relevant to the emergence of a new type of terrorism, just as the old terrorism that had threatened London for many years appeared finally to be waning. Financial districts in all parts of the world could usefully consider the City of London experiences in preparing for the uncertain future that the events of 9/11 have heralded.

Introduction

On 9th September 2001 ('9/11') the way in which terrorism is understood changed forever in the minds of many. The criminal attacks on New York and Washington DC represented the crossing of a new threshold in terrorism—demonstrating a potential for strategic change. The perpetrators of 9/11 sought to undermine the political, military and, crucially, economic life of the United States, symbolically in their targeting and practically in the scale of the destruction wrought. While the political and military prestige of the United States had previously been targeted by al-Qaeda abroad, in attacks against US embassies and military facilities, the targeting of New York's financial district as the symbol of American economic vitality, first attempted in the 1990s, may have represented a distinct attempt to undermine the economic basis of American power. In destroying the World Trade Centre towers and killing almost 3000 innocent people in their places of work, the terrorists fundamentally altered how Americans felt about their security and changed for ever the concept of a secure homeland in their minds. As well as the sheer human cost of the attacks, the financial implications of 9/11 appeared at times to be overwhelming.[2] Many predicted a global recession, as the already precarious economies of the developed world were pushed over the edge by the attacks, and consumer confidence, that essential motor of American growth, weakened. Statements from al-Qaeda also pointed to the economic costs of the attacks, suggesting that America was really a paper tiger economically. The true costs will take some time to emerge, although according to the OECD the direct costs of the attacks and the subsequent clean-up have been something over $25 billion.[3]

Terrorists had targeted the economic structures of the state before, but in the 1990s attacks in city centres and financial districts, in particular in both New York and the City of London, convinced business leaders that a new and dangerous stage in terrorist activity had arrived—one which required a new and dramatic response. The weapon of choice around the world for most terrorist groups remains the vehicle-borne conventional explosive device, and it remains highly effective and relatively straightforward to deploy. In Britain, the dangers posed by Irish Republican terrorism prompted the authorities in London's financial district to adopt extraordinary security measures to prevent and deter attacks, and, as significantly, to ensure business continuity in the event that the security measures failed. The London experience provides a number of useful lessons for counter-terrorism policies in advanced urban centres.

From the early years of the current Northern Ireland 'troubles' in the 1970s to the cease-fires of the mid-1990s, the British mainland provided important targets for Irish Republican campaigns. During this period a range of terrorist tactics were adopted, from assassination to the disruption of transport systems, from crude attacks on civilian areas to more sophisticated strikes against economic targets. The aim of this chapter is to describe aspects of the final campaign of 1988–96 on the mainland, to consider the strategic reasoning behind it, and to assess how the City of London managed the threat to its security. There is a danger of making a series of disparate actions appear to be more rational and focused than was the case. However, towards the end of this campaign the IRA clearly believed that they had identified a target set of great importance to the British establishment, and believed that this was an important factor in persuading the British government to take seriously a peace process which granted a leading role to its political wing, Sinn Féin.[4]

Targeting the mainland

At the start of the present 'troubles' the IRA leadership had been reluctant to start a mainland campaign because of past failures and the logistical and financial demands of establishing an effective operation. In the three serious bouts of activity on the mainland post-1945, the violence never approached the scale of that seen in Northern Ireland, despite horrific individual attacks. The worst year of violence on the mainland was 1974, when 46 people were killed,[5] most through bombings—notably the two Birmingham pub bombings, which killed 21, and the bombing of a coach, which killed 12. The second campaign, following the prison hunger strikes in 1981, culminated with the attempt to kill the prime minister, Margaret Thatcher, in the Brighton Grand Hotel bombing in 1984. The final campaign started in 1988, with an attack on a barracks in North London in which one soldier was killed and eight others injured. Thereafter there was a steadily escalating level of violence, reaching a peak in 1992 when six people were killed and 212 injured in attacks on the mainland. Between 1983 and 1993, according to one source, in the 20 worst atrocities committed by the IRA after it bombed the famous Harrods department store, 35 people died, 544 were injured, two attempts were made on the life of an incumbent prime minister and damage to property amounted to £1.5 billion—another £350 million was spent making military bases and government buildings more secure.[6] Nonetheless, in cold statistical terms, the actual effect of the mainland campaign was marginal compared to the slaughter in Northern Ireland. Between 1969 and 1998 the total number of deaths from the Northern Irish conflict in the United Kingdom was 3343, of which 125 were on the mainland.[7] The annual death rate from terrorism was falling towards the end of the campaign: in 1972, 465 people were killed in Northern Ireland, while in 1994 the figure was 61. By comparison, attacks on the mainland in the campaign that finished with the first cease-fire in 1994 led to a total of 28 deaths.

London was clearly the most favoured target for the IRA. In the six years after 1988, incidents in the capital totalled 103 out of 166 for the mainland. The IRA was responsible for all but three of these, and was the only serious threat following the penetration of the Irish National Liberation Army (INLA) and the capture and imprisonment of some of its most senior members in the early 1990s. Some parts of the country were barely touched by the IRA. Others, like the North-West and the North-East, suffered attacks, possibly because of the existence of particular cells based in their areas. The South-East, however, suffered continuously. Even many of the attacks outside of London were directed against London itself through the campaign against the transport system. In addition to transport and some military targets, London was also targeted for its importance as the political and economic centre of the United Kingdom. South London was barely attacked at all, while Oxford Street, the main shopping street, was a regular and favoured target. Between 1972 and 1996 some 50 people died in London in violence related to Northern Ireland.

By comparison, the number of attacks on the financial district, the City of London (or the 'Square Mile') was small. Between 1988 and 1994, it suffered six attacks, all bombings, which caused four deaths and 133 injuries. However, the attacks included two of the largest terrorist bombs ever detonated on the mainland, and caused hundreds of millions of pounds worth of damage. It was the size of the bombs used against the City, rather than their frequency, which gave rise to concerns. On 10th April 1992 there was a major explosion outside the Baltic Exchange in St Mary's Axe in the City, killing three and injuring 91. The anxiety in the City after this bomb, particularly in relation to insurance, appears to have persuaded the IRA that such attacks had an unusual potential to unnerve the British establishment. In the second half of 1992 the IRA regularly attempted to detonate large bombs against large economic targets, but was frustrated in the process. The IRA eventually succeeded in April 1993, with the Bishopsgate bomb in the City of London. In its propaganda at the time, the IRA focused on the various forms of economic damage such attacks caused.

IRA strategy and the final mainland campaign

By the late 1980s/early 1990s estimates of IRA activists ranged from 300 to 500, with another 700 second-rank supporters, and possibly a few thousand sympathisers.[8] From a structure of 'battalions' each employing perhaps 40, IRA organised itself into cells of no more than eight.[9] The IRA's political wing, Sinn Féin, had several thousand members, a proportion of whom were also believed to be in the IRA.[10] Estimates of the number of activists involved in the final mainland campaign of the late 1980s/early 1990s ranged from eight to 25. Operations on the mainland were allegedly controlled by the 'England Department', possibly by a mainland-based 'England Officer'.[11] On this, as on many other matters, there was little hard evidence.[12]

The IRA was armed with large quantities of Semtex explosives during this campaign. It preferred to be self-sufficient, and dealt directly with friendly states rather than other terrorist groups, using the Irish Republic or the United States as sanctuaries.[13] Weapons supplies were built up with Libyan help during the 1980s[14]—indeed, some have argued that it was the supply of Semtex in quantity that lay directly behind the escalation of the IRA's mainland campaign in the late 1980s.[15] It was also alleged that Iran become more important to the IRA towards the end of the third campaign.[16] The IRA was also in receipt of adequate funding—possibly several million pounds per year.[17] IRA strategy concentrated on undermining government within Northern Ireland, largely through attacks on people and facilities linked (sometimes quite loosely) with the security forces and also through attacks on commerce.[18] For the IRA to succeed, its military campaign needed to appear irresistible, yet for many years it did not come close to sustaining the necessary level of activity in Ireland. The fact that the IRA was caught between victory and defeat in its Northern Ireland campaign made its mainland campaign even more important:

> We can't be beaten: there is no question of us winning in the sense of driving the British Army into the sea. But we always maintain the capacity to bring the situation to a crisis at some stage.[19]

Little happened on the mainland following the Brighton bombing, although this does not seem to have been so much a strategic decision as the consequence of arrests and aborted campaigns. It may have been only with the arrival of large quantities of Semtex and other equipment from Libya in the late 1980s that the campaign, as a natural consequence, was again escalated.[20] This final campaign would achieve what the IRA had set out to do—prove that it could sustain a long-term mainland campaign. The first attack on a target in the City of London in this campaign came on 20th July 1990, when an IRA bomb exploded at the Stock Exchange building. Although the Exchange was closed for the day as a result of the explosion, trading was largely uninterrupted, since the central computer system remained undamaged. When the next spectacular came, the choice of target reflected an increasing belief in the importance of attacking Britain's economy rather than its political system. A growing interest in economic targets had been evident in Northern Ireland for some time, and after bombs badly damaged the Europa hotel and the Grand Opera House, the Northern Ireland Office announced a moratorium until April 1992 on capital spending, in order to meet the cost of attacks against property. The two bombs were estimated to have cost the British taxpayer £10 million in indemnity payments.[21] Insurance companies had stopped providing cover for bomb damage in the 1970s. Instead, from 1969 to the end of 1992, the Northern Ireland Office (a department of the UK government) found itself paying a total of £657 million for the damage caused by almost 10,000 explosions. In addition, payments to victims of criminal injury totalled a further £220 million.[22] It was later claimed that IRA planning for spectacular attacks in Britain began in 1991,[23] but the choice of target was probably influenced by this recent experience.

On 10th April 1992, a massive device exploded outside the Baltic Exchange in St Mary's Axe in the City of London, killing three and injuring 91. There was some speculation that the IRA had intended to bomb the Stock Exchange or Lloyd's. The bomb blast was the biggest ever seen on the mainland in peacetime. In the aftermath the IRA began to appreciate that they had touched a critical nerve ending. The republican newspaper (*An Phoblacht*, 'Republican News') picked up on the high initial estimates of cost (£1.5 billion) and the possibility that insurance firms might withdraw cover. It reported an IRA statement that an 'active service unit' had caused 'widespread damage to the commercial heart of Westminster [sic]'.[24] The next week the paper led on the campaign in England, stressing the inability of the UK authorities to do anything about it, pointing to leaked documents from the police which stressed their lack of intelligence:

> There are now claims of escalating costs to the insurance sector following the recent IRA bombing at the Baltic Exchange in the City of London which may be passed on to policy holders, and of the British exchequer having to pay compensation if insurance companies cannot cover war damage costs, as is the case in the Six Counties [ie Northern Ireland].[25]

The IRA were not incorrect about the scale of the damage—the financial cost of this one bomb, later estimated at over £800 million, was still more than the total cost of almost all the bomb attacks in Northern Ireland since the start of the troubles.[26] The main thrust of the campaign after this point was to recreate the effect of the St Mary's Axe Bomb. In the second half of 1992 the IRA was regularly frustrated in its attempts to detonate large bombs against large economic targets. Later, a senior police source was cited as saying that in this period at least ten unpublicised IRA attempts to bomb London had been thwarted. The City was not the only place where such large bombs could be detonated, but the concentration of high-value buildings and activities made it a more lucrative target. The City's small population, some 5000 residents, swells to 250–280,000 during a working day. Hence its vulnerability to disruption caused by attacks on the transport infrastructure, and a reduced risk that attacks at night or over weekends, using very large bombs, would cause the sort of massive loss of life that would be experienced in the more densely populated parts of London. In many ways the City was an ideal target.

By 1993 there were reasons to expect an intensification of the IRA campaign: a lack of progress in political talks; a presumed weakening in the position of the IRA's political wing, following the defeat of the Sinn Féin leader Gerry Adams in the April 1992 election; the presumed pro-nationalist sympathies of incoming US President Bill Clinton; and sufficient funds. On 24th April 1993, the biggest peacetime bomb in London exploded, devastating Bishopsgate, and with it City confidence. Unlike Semtex, which can be lethal even in small quantities, the ammonium nitrate fertiliser used in the City bomb needs to be gathered in considerable quantities, mixed with other base materials, and then packed into several containers; the explosion was huge. The IRA estimated that the cost would be high—a low figure of £350 million and, enthusiastically, a high of £2.5 billion. All the buildings and businesses affected by the blast were listed. *An Phoblacht* remarked on the property damage, removal expenses, the finding of alternative space, communication costs and business interruption they would all now suffer:

> As well as the huge costs of structural damage, the loss of business and the knock-on effects of insurance costs, the City of London is assessing the damage to its prestige as a world financial centre.[27]

Anxiety was said to be growing in the City as the government acknowledged that total security could not be guaranteed. BBC Radio interviews were cited. A Lloyds broker said: 'it would be silly not to have second thoughts about staying in the City area, and business is far from usual'. A banker from a large Japanese bank was 'very much frightened when I come to London'. A Czech banker claimed: 'Continuation of these attacks would lead people to move to Germany or the new financial centre in Dublin.'[28] The attitude of foreign bankers was crucial—as both sides recognised. Two weeks later *An Phoblacht* returned to the issue, now reporting with satisfaction that 157 buildings had been damaged, 25 severely and 30 moderately, and noting the number of banks and overseas security firms based in the City, with their annual contribution of £17 billion to the British economy.

During 1994 there were a number of almost routine attacks designed to hurt retailing and transport, before the next 'spectacular', a mortar attack on Heathrow airport. Although there were other attacks in the first half of 1994, the Heathrow attack was the last spectacular of this campaign before a cease-fire was announced in August 1994, when the IRA announced a 'complete cessation of hostilities', although without using the word 'permanent' as hoped for by the British government.[29]

The relationship between the campaign and political objectives can be illustrated by looking at IRA statements. After the St Mary's Axe bomb, it observed simply that:

> [T]he attacks are a direct consequence of Britain's illegal occupation of Irish territory and such attacks will continue so long as Britain persists in that occupation.[30]

The post-Bishopsgate communiqué, a year later, provided more of a sense of a political process:

> The leadership of the PIRA repeats its call for the British establishment to seize the opportunity and to take the steps needed for ending its futile and costly war in Ireland.
>
> We should emphasise that they should pursue the path to peace or resign themselves to the path of war.[31]

And then in July 1993, a few weeks later, in a notorious letter to Japanese banks (but not to American banks in the same buildings), the IRA observed:

> The British have the power but not yet the inclination to bring this conflict to an end. We in the PIRA point out that peace will only ensue when the causes of the conflict are removed. This can only be achieved through inclusive negotiations leading to a democratic settlement which recognises the fundamental and immutable right of the Irish people to national self-determination.[32]

While it made propaganda sense to strike a more conciliatory tone in this message, it is now known that communications between the British government and the IRA were in place at this time.

The City was thus not targeted in order to cause loss of life, which could readily be accomplished elsewhere, nor to disrupt transport, but to cause economic pain, thereby putting pressure on the government. After the Bishopsgate bomb, a *Financial Times*/MORI survey found only one third of organisations in the City of London either 'fairly' or 'very satisfied' with the performance of the government in dealing with the problems caused by the IRA, while 34 per cent were 'fairly' or 'very dissatisfied'. Notably, the most highlighted aspect for dissatisfaction was 'a perceived failure' by the government to change the political situation in Northern Ireland.

The City's response, the return to violence and the second cease-fire

The very fact that the City had been so clearly singled out for special attention led it to take measures to defend itself against further outrages. Indeed, the danger of the City's future as the leading financial centre of Europe being undermined had been identified immediately following the St Mary Axe bomb of 1992.[33] Planning on this had begun before the cease-fire with a round of consultation exercises with leading institutions in the City. The result was the 'ring of steel', adopted as a environmental traffic-calming measure in 1993, which gave the City of London extra powers to control the terrorist threat without involving central government. All vehicle access into and out of the City (admittedly only just over one square mile of a thinly populated part of London) began to be controlled by uniformed police and sophisticated electronic surveillance devices, at a number of entry and exit points which were reduced from 30 to eight. Plans for the total pedestrianisation of the City were held in reserve at this stage. Numerous roads were closed to traffic, and suspect vehicles stopped and searched. This enormous traffic management scheme was backed up by heavily armed police road blocks established at random points and times around the City.

There was concern in neighbouring areas that the City's measures would simply force the violence elsewhere, and this point was vindicated in 1996 when the IRA cease-fire was suddenly broken by an attack on the Canary Wharf building in East London. On 9th February the IRA planted a huge bomb in the Docklands financial centre (thus technically outside the City of London itself), killing two and injuring more than 100. Damage was estimated at £85–150 million. The London Docklands Development Corporation, which owned the Canary Wharf site, thereafter instituted measures similar to those in the City of London itself, including checkpoints and sensors. One complicating factor in the case of Docklands was a significantly larger resident population of over 25,000, compared with the City's 5000.

These attacks indicated that the IRA was convinced of the importance of these spectacular attacks, even if the evidence for their effect on the British government's decision-making was somewhat shaky.[34] While increases in violence had always concentrated the minds of ministers, and the St Mary Axe bombing had certainly come as a shock, the Bishopsgate bombing may have been a retrograde step for republicans; some have even characterised it as the myth of the Bishopsgate bomb. The back-channel contacts between the British government and the republicans which had been established endured through the St Mary Axe bombing, but following the Bishopsgate bomb the British emphasised that 'events on the ground' were affecting attitudes, a reference to the attack on the City. The prime minister, John Major, may have come to doubt the ability of moderate republicans to deliver the movement into the democratic political arena. The result was effectively to close the back-channel.[35] At Canary Wharf, when the IRA wanted to send a message to the British government, it was London's financial 'soft underbelly' that was chosen again. John Major asserts that the attack was later revealed to have been planned three months in advance, ie during the cease-fire.[36] He significantly did nothing other than state that no ministers would meet with Sinn Féin again until another meaningful cease-fire was in place, arguing that he had gone too far down the road to peace to play into the hands of those against it by undertaking more vigorous measures in response. Another interpretation is that the IRA use of violence had achieved its aim.

> [T]he violence exerted a tremendous coercive pressure on the British and Irish governments to renew efforts to bring Sinn Fein and the IRA back on board with concessions, most notably by dropping the provisions for decommissioning terrorist weapons and the promise of early entry into negotiations on condition of a renewed cease-fire.[37]

One way in which IRA may have struck a nerve was the continuing question of insurance. Until 1992 insurance against terrorist attack in Britain appeared as a relatively small part of the cover provided by insurers. The St Mary's Axe bomb was seen to represent a third of the total of around £2 billion of premiums paid by industry each year for protection against all other forms of risk.[38] The bulk of the costs of the Baltic Exchange incident fell on the reinsurance companies, such as Munich Re, who had insured the insurers (and which then, with Swiss Re, was among the first to pull out of giving such cover). In May 1993, the City of London police commissioner, Owen Kelly, criticised the insurance industry for disclosing that the St Mary's Axe bomb's cost of £800 million (later revised down to £500 million) compared unfavourably with a total of £657 million paid out over more than 20 years of terrorism in Northern Ireland:

> A clearly unintended consequence of that statement is the increased threat level we now face from the terrorists having had that comparison and its implications pointed out so publicly.[39]

After Canary Wharf the pressure was maintained in a series of attacks on London, but with no loss of life excepting the death of an IRA activist, killed by his own bomb in February 1996 while travelling on a bus towards the City, injuring eight other passengers. The strategy of focusing on the economic base appeared to have started to unravel by this point, although one more highly significant attack almost occurred. In the early summer of 1996, the IRA's 'England Department' sent over a seven-member active service unit specially chosen for an elaborate series of attacks on the mainland. In July all seven were caught in a stunning series of successes for the security services, which revealed the extent of the plan. It was for a series of attacks on electricity control stations, with the aim of shutting down the whole of the City of London's electricity supplies in a co-ordinated attack, blacking-out the whole of London if possible. The continuing importance of the City of London to the IRA was revealed at the trial, when one of the unit's members claimed that the attacks were designed merely to be hoaxes that would reveal the futility of the City of London's 'ring of steel' security scheme.

> If the IRA were capable of closing down all electricity in London without going into London, it would make the ring of steel null and void.[40]

Claiming that the devices were not intended to explode, Gerard Hanratty stated that the 37 boxes containing electrical timing devices were to be planted in electricity sub-stations outside London—in Amersham, Elstree, Waltham Cross, Canterbury, West Weybridge and Rayleigh—but would cause the entire electricity grid to be shut down simply for bomb disposal officers to check to see if the devices were live or not. The unit hoped for disruption to last as much as a day and a half.[41] The prosecution claimed that the plot was in fact a real one, and that if the 37 devices had exploded the impact would have been to disrupt London's electricity supplies for 'months'.[42] The dislocation of economic activity was potentially huge, and would have been a great embarrassment to the government. The significance of such attacks was highlighted after 9/11, when the director of the FBI's National Infrastructure Protection Centre stated: 'The event I most fear is a physical attack in conjunction with a successful cyber-attack on the ... power grid'.[43]

In September 1996 the security forces broke up another IRA active service unit in a spectacular series of arrests; these followed an enormous surveillance operation which uncovered over six tons of explosives, Semtex, handguns and bomb-making equipment, including four bombs primed and ready for use. Thereafter it is believed by some that the IRA may have closed down its 'England Department', fearful of penetration by the security services.[44] Following the British general election of May 1997 the mood improved, and the republicans had high hopes of Tony Blair's new Labour Party government. The stage was set for the resumption of the cease-fire, which duly occurred on 20th July 1997. Despite splits in the republican movement, the splinter groups were unable to sustain successful operations on the mainland, and the peace process moved decisively forward in April 1998 with the Good Friday (or Belfast) Agreement.

Assessing the City's response

The challenge for the City had been to avoid a withdrawal of foreign financial institutions and domestic firms, and it set out to do this by preventing another major explosion and ensuring that its disaster recovery systems minimised the dislocation of business activity if its defences were breached. These objectives were complementary, since a vital plank of counter-terrorist policy was to demonstrate the limited impact another bomb would have and so reduce the attractiveness of the City as a target. The problem for the City's planners was that the greater the effort to prevent another bomb attack, the greater the challenge to IRA to breach any security measures.

The 'traffic management scheme' was designed to monitor all traffic entering the square mile through blocking-off streets and establishing manned police checkpoints. All vehicle traffic entering the City had to pass police cordons where random searches took place, primarily of larger commercial vehicles. The scheme did not claim to be foolproof, not least as pedestrian and railway/underground traffic was not affected by the measures, but emphasised the danger of large vehicle-borne bombs. In addition to the police presence at entry points, high-definition closed circuit television cameras were placed at all entrances and at strategic sites around the City. Vehicle number plates were electronically fed into databases as they passed the checkpoints and checked. Private companies were encouraged to join a 'camera watch' scheme whereby existing security cameras were directed to cover the streets outside buildings, increasing the risk of discovery for the terrorists. Backing up the static security measures were mobile 'rolling road blocks' of armed City of London Police units which established checkpoints at random, checking vehicles and occasionally pedestrians. These proved to be highly effective in terms of reassurance and of economy (the teams involved no more than 12 police officers), and heightened the danger for IRA activitists. Over 1000 litterbins were removed, and the amount of time rubbish bags could be left on the streets was reduced.

Disaster recovery measures were also taken, both at Corporation (the City's local government body) and individual level. A pager alert system was introduced by the City to keep businesses informed about potential and actual threats, allowing for better co-ordination of disaster recovery. Companies were encouraged to develop disaster recovery plans, establishing back-up facilities where, for example, foreign exchange dealing rooms could be relocated. Such measures were relevant to all emergencies, not just terrorist attacks, and by 1994 the majority of all firms operating in the City had some form of contingency plan. Opinion surveys found a large majority of those who used the City, mostly for work, supported the measures as the necessary price for security, but it is important to note that the resident population of the City was small compared to the number who commuted in and out of it daily, making the question of civil liberties less pressing for those surveyed. The City of London Police were delighted to report a significant drop in all crime following the scheme's introduction, but this was a two-edged sword—the drop in crime resulted from measures of control that few would have liked to see extended.

There were initial objections to the traffic management scheme from neighbouring boroughs and motoring organisations, which feared an increase in traffic and pollution and possibly terrorism in their districts. Some politicians expressed concern that the measures gave a propaganda victory to IRA, encouraging it to target the City. Such concerns were not without force, and in the short term the scheme did indicate to the IRA that it had hit upon a particular vulnerability within the British state.

For the Corporation, however, the scheme proved its worth—no major bomb attacks occurred within the City following its implementation. When the IRA broke its cease-fire in February 1996 with a 1000lb vehicle bomb, targeting high-cost business premises as before, the bomb was detonated outside the City, in Docklands. For the Corporation it was a crucial vindication of its actions. To critics, it suggested that the terrorist threat had been exported to the City's neighbours. In the City's defence it could be noted that the Docklands business district had actually been targeted before by the IRA.

Now the traffic management scheme has become part of London life. Indeed, it is not possible to contemplate removal of the current traffic restrictions, and the scheme has been extended twice to take in additional office developments.[45] The major problem with regard to sustainability lies in the demands placed on police manpower. Much of the surveillance work by necessity involves long hours and is boring by nature. Other police activities inevitably suffer unless there is either a significant increase in manpower or a shift of the burden of responsibility for some related duties to other personnel. Alternative forms of manpower may well lack police training and discipline, but at some point consideration may have to be given to enhanced use of private security guards.

In implementing its security measures the City benefited from a number of unique circumstances which set it apart from other districts of London. Most significantly, it possessed its own police force, the City of London Police, one of the country's smallest forces with less than 1000 officers. This enabled the Corporation to mobilise its own police resources in a manner not open to a normal London borough, which could only call upon the London-wide Metropolitan Police force and thus faced competing demands for scarce police resources. Canary Wharf, by contrast, was not able to call on additional police resources except in times of heightened alert, and has had to opt for the private security route. The unusual structure of the Corporation, which includes appointed representatives from the business community and only a minority of elected representatives, also allowed consensus to be reached quickly on the security measures.[46] Furthermore, the public's acceptance of the scheme had been aided by the unusually low resident population of the City.

Conclusion

The events of 1988–96 saw the IRA edge its way towards a discernible strategy of economic targeting, although it was not pursued single-mindedly at all times. The City attacks do appear, however, to have convinced the IRA leadership of the effectiveness of pressuring the economic base of the British state in pursuit of political concessions. Large vehicle-borne devices placed in high-density urban areas of expensive commercial property afforded highly effective devices for exploiting a vulnerable flank in the British state. Furthermore, such 'spectaculars' represented a more attractive option than so-called 'information age' attacks or cyber warfare (which might have disrupted trading activity just as well). This was because they fitted in with the organisation's self-image and tended to generate a type of press coverage that few other types of attack could. Tower blocks with a thousand windows blown out would make for photographic and telegenic images—dealers with their heads in their hands would not offer the same impact visually. The St Mary's Axe bombing revealed the sensitivity of the British government to attacks on its financial centre, and Bishopsgate confirmed this to the IRA. Canary Wharf demonstrated their understanding of the importance of this strategy. However, the plan to target the London electricity network to black-out the City revealed the organisation's willingness to be flexible in how it disrupted economic activity.

The City of London meanwhile was genuinely shocked by these devastating attacks, and resolved to use its unique corporate structure, financial muscle and political influence to counter the threat. The solutions it settled on were not necessarily approved of by the British government, but the City's autonomy and unique importance to the whole British economy forced the government to acquiesce. Despite denials to the contrary, the IRA logic behind these attacks may also have struck a chord with the British government. It is significant that following the City and Canary Wharf bombs, the government continued to attempt to engage the terrorists in dialogue. The fact that the insurance costs of the two City bombs were greater than the entire insurance bill from all preceding Northern Ireland bombings cannot be ignored, and doubtless was not. However, the attacks were also timed to avoid significant loss of life, making the possibility of negotiation with the perpetrators politically possible in a way which would have been impossible with those behind 9/11. Nevertheless, the IRA did not by any means achieve all their political objectives.

Given the City's unique circumstances, the extent to which lessons can be drawn from its experience is open to discussion. Clearly the threat of bomb attacks was not completely eliminated, but the danger of large vehicle-borne devices did appear to be substantially reduced as a consequence of the traffic management scheme. The major problem in implementing the security measures was the drain of police manpower away from other duties, and the sustainability of such a scheme in the long term remained in doubt. It may also be difficult to expect the general public to keep up their guard indefinitely. Furthermore, after a prolonged period without a major incident, persuading organisations to implement measures that seem expensive, or to keep their emergency plans up to date, will be harder. Finally, while the City's response in reality did displace the terrorist threat, it can also be asserted that there are only so many uniquely attractive targets in modern cities, and the dispersal theory should not be taken too far. Steps taken by the New York City authorities following the 1993 attack on the World Trade Centre and the Oklahoma City bombing of 1996 could have meant that when the attack came on 9/11 it was from the air and not from the ground.

In terms of disaster contingency planning, though, the City pointed the way forward for modern business districts, and specifically financial centres, where the dislocation of business needs to be counted in hours not days before the costs become prohibitive. Only through co-ordinated planning between public and private authorities could contingencies be effectively met and beaten. The experience of IRA terrorism made the City better able than its international counterparts to be confident about its capabilities in this area, and no significant relocation of foreign institutions from London occurred. The scheme thus offered benefits for all similar districts in terms of the good practice that it had demonstrated. Indeed, in terms of lessons for other centres, the City of London experience demonstrates that disaster recovery is as important as avoidance, not least as no security system is perfect. Only through detailed and co-ordinated advance planning can financial centres have any confidence in recovering from a major attack. Nonetheless, the basic principle remains that it is unwise to introduce measures beyond those that can realistically be sustained. Reducing an alert status can be as hazardous as an increase.

The London experience also points to the unique attractiveness for terrorists of the economic base of a country's infrastructure, and to the potential for spectacular (and photogenic) blast effect in city centres. Low-tech weaponry is revealed to be highly effective in what is fundamentally an asymmetric attack on an advanced Western state in the information age. In preparing for the threat of modern terrorism, 'old' terrorism and its methods should not be overlooked. When it finally arrived, the 'new' terrorism that was revealed on 11th September 2001 was not at all the sort that had been predicted. Instead of technologically sophisticated weapons of mass destruction, the perpetrators of 9/11 utilised the long-standing terrorist approach of careful planning, simple tactics and operational surprise to effect the most stunning terrorist 'spectacular' in history. The vulnerability of modern cities and of workplaces at their centre that has been revealed by the events of the last dozen years has demonstrated the need for a closer partnership between the public and private sectors in meeting these challenges, and the danger of complacency in disaster and continuity planning.

Notes

1. John Gearson is Senior Lecturer in Defence Studies, King's College London; email: john.gearson@kcl.ac.uk. He is currently seconded to the House of Commons as Committee Specialist of the Defence Select Committee. Following the Bishopsgate bomb attack he acted as a consultant, with Lawrence Freedman, to the City of London Corporation on the terrorist threat to the City.

2. Some estimates went as high as $95 billion, although how such figures were arrived at remained opaque. See BBC on-line discussion at *http://news.bbc.co.uk/1/hi/in_depth/world/2002/september_11_one_year_on/2239280.stm.*

3. *OECD Economic Outlook* (2002) The Economic Consequences of September 11th. No. 71, June.

4. For background on the IRA, see also Coogan, T.P. (1995) *The IRA*. London: Harper Collins; Bell, J.B. (1979) *The Secret Army: The IRA 1916–1979*. Dublin: Academy Press; Bishop, P. and Mallie, E. (1994) *The Provisional IRA*. London: Corgi. On the last 25 years of the 'troubles', see also Bell, J.B. (1993) *The Irish Troubles: A Generation of Violence 1967–1992*. Dublin: Gill and Macmillan.

5. Sutton, M. (1994) *An Index of Deaths from the Conflict in Ireland.* Cain Project website, at *http://cain.ulst.ac.uk*. Clutterbuck puts the figure at 40, revealing one of the problems with different counting rules; see Clutterbuck, R. (1993) Terrorism in Britain, in Wilkinson, P. (ed.) *Terrorism: British Perspectives.* Aldershot: Dartmouth, p 45.

6. Adams, J. (1993) Bungled: The Battle Against the IRA. *The Sunday Times*, 7th November.

7. This figure excludes deaths in the Republic of Ireland (107) and in continental Europe (18); see Sutton, op cit. Another authoritative and moving source on the

casualties of the troubles is McKittrick, D., Kelters, S. Fenney, B. and Thornton, C. (1999) *Lost Lives: The Stories of the Men and Women and Children who Died as a Result of the Northern Ireland Troubles*. London: Mainstream; this puts the total number of deaths from 1966 to 1999 at 3636, including 124 on the mainland. It is significant to note that the 'loyalist' paramilitaries killed more people in Northern Ireland in 1992, 1993 and 1994 than did the IRA; see *The Guardian* 1st January 1995.

8 Urban, M. (1992) Big Boy's Rules: *The Secret Struggle Against the IRA*. London: Faber & Faber, p 31, quoting Bishop and Mallie, op cit.

9 McGartland, M. (1997) *Fifty Dead Men Walking*. London: Blake, p 178.

10 McKittrick, D. (1992) *The Independent on Sunday*, 22nd November; *The Times* 17th December 1993. See also US Department of State (1993) *Patterns of Global Terrorism* 1992. Washington, DC: Office of the Co-ordinator for Counter-terrorism, Department of State, pp 48–9. *The Sunday Telegraph* (28th March 1993) put the figure at 800 hardcore activists, who could be relied upon to give physical support and shelter, and up to 10,000 sympathisers, many of whom would give active support. In 1994, world-wide, there were more than 1100 IRA members in jail (substantially more than the number of activists), more than a hundred of them serving life sentences for murder, and the bulk of them held in Northern Ireland. In all, some 350 IRA members were killed between 1969 and 1993; see *The Times*, 17th December 1993.

11 Sean O'Callaghan claims that between 1973 and 1979 Brian Kennan was in charge of this 'England Department'; see *The Sunday Times*, 13th November 1994.

12 One authority claimed that at any one time one or two ASUs, each comprising some eight to 12 people, were thought to have been at large on the mainland. The hard core element of an ASU may often have comprised no more than three people; see Clutterbuck, op cit, pp 45–6. One November 1992 estimate put the number at about 25, apparently based on a calculation of the number needed to sustain the campaign at its then level; see McKittrick, op cit. Another estimate at the same time spoke of 'at least a dozen IRA activists ... identified and connected to the present campaign'; see *The Times*, 17th November 1992.

13 On IRA links with outside groups, see McKinley, M. (1989) The Irish Republican Army and Terror International: An Inquiry into the Material Aspects of the First Fifteen Years. In Wilkinson, P. and Alisdair M. (eds) *Contemporary Research on Terrorism*. Aberdeen: Aberdeen University Press.

14 In 1985–86 Colonel Ghadaffi probably provided £2 million, in two payments, and possibly as much as 140 tons of weapons, including large quantities of Semtex explosives. The largest shipment of all was captured off the French coast aboard the *Eksund* in October 1987; see O'Brien, B. (1999) *The Long War: The IRA and Sinn Fein*. Dublin: O'Brien, p 129.

15 Sean O'Callaghan (1994) *The Sunday Times*, 13th November.

16 *The Times*, 29th April 1994; *The Financial Times*, 30th April 1994.

17 Including, it is claimed, £2 million from Syria in 1980; see O'Callaghan, S. (1998) *The Informer*. London: Bantam Press, p 103. McGartland (op cit, p 182) claims a figure of $10–12 million per year.

18 Remarkably little academic literature addresses the problem of Irish terrorism as a strategic phenomenon, one notable exception being Smith, M.L.R. (1997) *Fighting for Ireland: The Military Strategy of the Irish Republican Movement*. London: Routledge. See also Bell, J.B. (1990) An Irish War: The IRA's Armed Struggle, 1969–1990. *Small Wars & Insurgencies*. Vol. 1, No. 3. See also Bell, J.B. (1990) *IRA Tactics and Targets: An Analysis of Aspects of the Armed Struggle 1969-1989*. Dublin: Poolbeg, p 28, in which Bell expresses scepticism as to the validity of the term 'IRA strategy' as such: 'Almost no-one in operations or in authority within the IRA spends much time discussing matters beyond techniques, the potential level of intensity, and perceived vulnerabilities – and such matters almost never engender dispute, much less a vote in the Army Council. Megastrategy is for academics or military schools.' The lack of strategic analysis of the troubles is discussed in Smith, M.L.R. (1999) The Intellectual Internment of a Conflict: The Forgotten War in Northern Ireland. *International Affairs*. Vol. 75, No. 1, pp 77–97.

19 IRA Army Council spokesman speaking in 1986, quoted in Coogan, op cit, p 516.

20 O'Callaghan, op cit.

21 *The Financial Times*, 10th January 1992.

22 Ibid, 7th January 1992; 9th December 1992.

23 *The Sunday Times*, 28th February 1993.

24 *An Phoblacht*, 16th April 1992.

25 Ibid, 23rd April 1992.

26 Seldon, A. (1997) *Major: A Political Life*. London: Phoenix, p 413.

27 *An Phoblacht*, 29th April 1993.

28 On 29th April 1993.

29 In December 1993 the Prime Ministers of Britain and of the Irish Republic, John Major and Albert Reynolds respectively, had agreed an initiative designed to set a framework for peace in Northern Ireland. Thereafter the British Government expressed itself concerned over the dangers of allowing itself to be strung along into what were in effect negotiations in the name of clarifications. Following the 'Downing Street Declaration' there was no pause in IRA activity on the mainland, nor in Northern Ireland.

30 *An Phoblacht*, 16th April 1992.

31 Ibid, 29th April 1993.

32 Ibid, 8th July 1993.

33 Seldon, op cit, p 413.

34 Patterson, H. (1997) *The Politics of Illusion: A Political History of the IRA*. London: Serif, pp 230–1, quoting *Fortnight*, September 1990.

35 Ibid, pp 243–7.

36 Major, J. (1999) *The Autobiography*. London: HarperCollins, p 489.

37 Smith, M.L.R. (1998) Peace in Ulster? A Warning from History. *Jane's Intelligence Review*. Vol. 10, No. 7, July, p 6.

38 *The Independent*, 9th December 1992. The paper noted that comparable claims in Belfast were far lower. Thus Belfast's largest building, Windsor House, had been damaged twice, with claims the first time under £1 million and the second around that figure.

39 *The Daily Telegraph*, 7th May 1993.

40 Ibid, 5th June 1997.

41 Ibid.

42 Ibid, 12th April 1997.

43 *The Washington Post*, 27th June 2002.

44 *The Daily Telegraph*, 8th February 1997.

45 City of London Corporation (2000) *Broadgate Traffic Management Scheme*. Press release, 6th April.

46 In some ways the City is similar to private developments such as Battery Park City in New York.

Chapter 14

The Impact of September 11th on the UK Business Community

Bruce George, Mark Button and Natalie Whatford[1]

This chapter discusses the impact of the events of September 11th on the UK's business community, and the threat from what has been termed the 'new terrorism'. It summarises and analyses the effect on different sectors of the community, ranging from the public utilities to the financial and insurance sectors, and examines how security has been affected by the impact of September 11th. It also makes reference to the British government's response to the threat of the new terrorism, and to the House of Commons Defence Select Committee's reports The Threat from Terrorism *and* Defence and Security in the UK. *Finally, the chapter seeks to outline a number of considerations on how the UK could be better prepared for terrorist attacks.*

Introduction

On 11th September 2001 the most destructive terrorist attacks ever occurred when two hijacked planes were flown into the twin towers of the World Trade Center, a third smashed into the Pentagon in Washington DC and a fourth crashed in the countryside in Pennsylvania (probably on course for Washington DC). These attacks were perpetrated by al-Qaeda, and they have marked what could be described as a 'new terrorism' as they differ significantly from the aims and strategies of 'traditional' terrorists.[2] There are many legal and academic definitions of different forms of terrorism, which there is not the space to consider here.[3] However, some of the characteristics that distinguish 'new' from 'traditional' terrorism include:

- a determination to inflict mass casualties on innocent civilians;
- a willingness on the part of the perpetrators to kill themselves as well as their victims during an attack;
- an increased threat of the use of weapons of mass destruction, such as chemical, biological, nuclear or radiological (CBNR) weapons; and

- the pursuit of radical, opaque, non-negotiable objectives.

While examples of some of these characteristics can be found amongst 'traditional' terrorists, what is new is the deadly concoction of them all combined. An important question is whether 11th September was a unique event that is unlikely to happen again, or whether it is a benchmark which al-Qaeda or another group will seek to surpass in future attacks. The US 'war on terrorism', joined by many other countries in Afghanistan, has done much to disrupt and destroy al-Qaeda, but it still remains a very potent threat. As the House of Commons Defence Committee concluded:

> ... we can see no reason to dissent from the general view of our witnesses, and others with whom we have discussed these issues, that there is a continuing threat to UK interests posed by the existence of organisations or groups whose aim is to inflict mass casualties.[4]

It has therefore been prudent for the British government to pursue initiatives to reduce the risk of such an event ever occurring, or to deal with the consequences should one arise. It is not just the government that has had to respond to the impact of the events of 11th September: they have also impacted on many other public-sector organisations and on the business community. It is the aim of this chapter to examine the latter in the UK, focusing on the security-related issues, although the implications clearly extend beyond security. Before we can undertake this, however, it is important to examine briefly what the government has done, as this also impacts on the business community.

The UK government's response to 11th September

The United Kingdom has endured terrorist action for decades, from both indigenous and international groups, though that experience and its relatively successful response should never imbue a sense of complacency. This experience has led to the development of a wide range of strategies (though it must be pointed out that non-terrorist activities, notably the foot and mouth epidemic, the fuel protests and the 'Y2K' preparations have all contributed to an awareness of the need for organisational change).

The UK both at home and in its interests abroad has experienced terrorism from 'ideological' terrorists, 'religio-political' groups and 'single-issue' groups (notably animal rights activists), as well as 'state-sponsored' terrorists. The most significant experience has been with 'nationalist' type terrorism[5] in Northern Ireland, in mainland UK and abroad. This has encompassed a wide range of attacks, from the shooting of civilians and military or security personnel to the detonation of large bombs in the City of London. The tactics of the Provisional Irish Republican Army (PIRA) during the early 1990s, when they exploded large bombs in key economic areas such as St Mary's Axe, Bishopsgate, Canary Wharf and Manchester, causing civilian casualties and tens of millions of pounds worth of damage, has led to the development of a wide range of strategies by the government and other organisations.[6]

The events of 11th September posed new challenges, and the government has done a great deal to strengthen the UK against the threat from terrorism by enhancing the ability to respond should such an attack occur. As the House of Commons Defence Committee reported, the government has been engaged first in enhancing protection and prevention and second in improving consequence management and resilience, which the Committee's report described as:

> the ability to respond to any successful attack in a way that minimises its effects and ensures as far as possible the continuity of government, other public agencies and society as a whole.[7]

The UK government has participated in the 'war against terrorism' in Afghanistan, collaborated with international organisations such as the UN, NATO, the EU, OSCE, the Council of Europe, G7, G8, various inter-parliamentary assemblies, Interpol and others in strengthening international and national capabilities. It has enhanced its intelligence capabilities after what was perceived as the failure to anticipate 11th September. It has passed legislation such as the Anti-Terrorism Crime and Security Act 2001, which closely followed the Terrorism Act 2000.

There have been reorganisations and reviews within government to take into account this new threat, particularly after 11th September; these are described in some detail in the Defence Committee's report, including the establishment and enhancement of the Civil Contingencies Secretariat actually set up prior to 11th September. There is also an important role for the Civil Contingencies Committee (CCC), which is chaired by the Home Secretary. Its task is:

> to co-ordinate the preparation of plans for ensuring in an emergency the supplies and services essential to the life of the country; to keep those plans under regular review and to supervise their prompt and effective implementation in specific emergencies.

In order to fulfil its task the CCC has three sub-committees; one on chemical, biological, radiological and nuclear consequence management, another on London resilience, and a third on UK resilience.[8]

Central government plays a crucial role in seeking to protect society from the consequences of acts of terrorism. It passes legislation and issues regulations, sets standards and guidelines, and establishes regulatory regimes within which the private sector should or may improve its own security. Its armed forces are deployed to counter terrorism at its source, and it has a near monopoly role in defending air space and territorial waters. Its military personnel, police and others defend military and other key sites, although the process of privatisation has transferred much of what may be called the critical national infrastructure, such as nuclear installations, ports, aviation and transport, into the private sector.

The Ministry of Defence (MoD) has embarked on the publication of a so-called 'New Chapter' of the 1998 Strategic Defence Review, and given an additional role to the armed forces reserves. This will ensure that it has the right concepts, forces and capabilities to meet the additional challenges that international terrorism and asymmetric threats bring. Its initial thinking is contained in a Public Discussion Paper that was published in the summer of 2002. The military would be engaged in well-rehearsed roles variously designated 'military assistance to the civil authority' (MACA), 'military aid to the civil community' (MACC), 'military aid to the civil power' (MACP) and 'military aid to other government departments' (MAGD).

The police forces in Great Britain and Northern Ireland have been at the forefront in dealing with terrorist organisations, in both prevention and investigation. There has been a high degree of coordination, with a crucial role for the Anti-Terrorist National Coordinator. The Metropolitan Police has increased its anti-terrorist activities, and the Association of Chief Police Officers provides the forum within which senior police and others discuss the issues. A special mention must be made of the Assistant Commissioner (Specialist Operations), David Veness, who has not only played a vital role over the years but has written and spoken very extensively in co-ordinating with non-police bodies, including the business community.

In 1998 the police established a 'National Terrorist Crime and Prevention Unit' (NaCTSO) whose purpose is *inter alia* to achieve a cohesive national approach in the provision of advice on counter-terrorist protective security and on related community safety issues. It also seeks to promote terrorism prevention within the broader national policing plan. Officers from Scotland Yard and the City of London Police have been in regular discussions with the business community, particularly senior executives from London-based companies in the City, Canary Wharf and the West End, giving advice on how such companies should be conducting business and setting the necessary security preparations in place. What is obvious is the vital contribution that has to be made by the commercial sector, by the private security industry and by what the Defence Committee called 'constructive public involvement'.

Our studies reached the tentative conclusion that those sectors of industry and commerce that had already responded positively to the threats of terrorism were the most likely to have reappraised and improved their capabilities following 11th

September. These sectors include the financial services, aviation and aerospace industries, transport, the nuclear industry and its installations, communications, the water industry, areas of sport and recreation, and the retail sector. Some areas have been identified as particularly vulnerable, such as the City of London and Canary Wharf. This list is not exhaustive and there are sections even of the above where differential standards have been set or aspired to. The willingness to respond positively may be linked to vulnerability, to pressure from government or insurers, to a competent security manager or to historical experiences of terrorist acts. What is truly amazing, however, is how many commercial enterprises have taken little or no action. This may have come about for a number of reasons, such as apathy and indifference, or such perceptions 'it couldn't happen to us', 'we're unlikely to be a target', 'it's the Government's responsibility', 'our insurance will cover us' and 'we can't afford to do any more', to name some of the excuses given.

The UK business community's response

> On the basis of statistical analysis of trends in targeting by international terrorist groups over recent years, it is not difficult to predict the most likely targets in coming years. Over half the attacks on property are likely to involve business or industry premises, roughly 10 per cent are likely to involve diplomatic premises and about half this number will involve other government premises and military facilities.[9]

The above statement by Professor Paul Wilkinson, and the nature of the 11th September attacks, illustrate why the UK business community needs to be prepared for terrorism. Further evidence from the US Department of State shows that of 531 facilities attacked in 2001, 397 were businesses.[10] The term 'UK business community' could embrace a very wide range of organisations, and it is not the purpose of this chapter to get too embroiled in definitions. Therefore we will interpret the term broadly in examining the impact of 11th September on private sector or commercial organisations. The focus will be on security-related issues, although there are clearly many other implications.[11] The discussion will also be of a general nature rather than focusing on specific organisations, in order to avoid revealing sensitive information, although there will be some discussion of specific sectors such as aviation because of the much more significant impact upon them. All references to the names of companies and organisations have also been removed, to protect their identity.

The nature of the 11th September attacks illustrate a wide range of potential targets for the future, many of which are located in the business community. The aviation sector is the most prominent because of the lethal use to which aircraft can potentially be put, along with the huge publicity such attacks invoke. Similarly, ports and ships have been identified as another potential target for terrorists.[12] The critical infrastructure of the country is also at risk, of for instance the sabotage of telecommunications or the poisoning of water supply networks.[13] There has also been concern raised at the potential for an attack on a nuclear installation,[14] or on large buildings (particularly tall and highly symbolic structures). Linked to the identification of such possible targets has also been the realisation that al-Qaeda's desire to maximise casualties may involve an attack on one of the above with CBNR weapons.[15]

The pursuit of privatisation by the Conservative and Labour governments over recent years, however, has meant many of these potential targets are now located in the private sector. The key networks of electricity, gas, water, telecommunications, broadcasting, railways and ports are all now largely owned and operated by the private sector; significant parts of the nuclear industry have also been privatised. These are all prime targets for terrorists. The MoD has for a long time maintained a list of 'key points' which are essential to the 'ability of the country and the armed forces to conduct military operations', and the Security Service a list of economic key points which are part of the critical national infrastructure. Arrangements have been made for the protection of these key points to be supported by the armed forces in the wake of a terrorist attack.

Through a myriad of legislation and other means in these areas—and other vulnerable sectors—there is scope for the government to mandate specific security strategies.[16] It is in the field of transport security that government has intervened most and introduced most changes. On 11th September the highest level of security possible was introduced, and on the following day the Civil Aviation Authority published a requirement for aircraft cockpit doors to be locked. A day later TRANSEC (the government organisation responsible for regulating transport security) published a letter further increasing security measures for the short term. Measures have also been mandated to tighten restricted zones, such that all items, including items for sale, must be security-checked. Over the longer term the government has instituted a review of airport security, under the chairmanship of Sir John Wheeler. Airports and airlines have also done much themselves, such as investigating the introduction of new screening equipment, reviewing the allocation of airport passes and investing in new CCTV systems. The threat of a seaborne terrorist attack, where a ship brings in terrorists with weapons of mass destruction, has also been raised. In the UK alone some five million containers are brought into the UK each year, and only a tiny fraction of these are searched. In the USA more thorough searches and the placement of US personnel abroad to carry them out have been imposed. The UK has not gone this far, but it has put in train a greater number of searches.[17]

It is also a responsibility of an employer under the Health and Safety at Work Act 1974 to protect from harm employees and visitors to the organisation's premises. Clearly, the hazards posed by an incident of 'new terrorism' could cause significant injuries and deaths. If an organisation—particularly in a vulnerable location—has not established new procedures and strategies to deal with such an attack it could be at risk of litigation should one occur. In the light of this, prudent organisations have undertaken reviews of and changes to a wide range of operational areas. Some of the more specific examples will now be explored.

Introduction of new security strategies

The security strategies of an organisation emerge because of two significant sets of influences, as illustrated in Figure 1.[18] The primary influence is the nature, origin and potential consequences of risks. Thus the risk of aeroplanes been hijacked in the past has led, along with a variety of other strategies, to screening and searching procedures at airports to deter and detect potential weapons that could be used by hijackers. The government—or a body an organisation is a member of—may also impose or suggest security strategies that an organisation then pursues. The example cited above of TRANSEC imposing tighter screening procedures would be an example of this. In developing a security strategy to deal with a terrorist attack, one of the problems for many decision makers is that, because they do not have access to sufficient information on the risks from terrorists, their risk assessments, which should be the foundation of any security strategy, may not be appropriate. Hence many security decision makers have been demanding greater access to information on the risks; this will be considered later in this chapter.

Figure 1. A model of the security decision-making process

Nature, origin and consequences of risks

<>

Any statutory/voluntary requirements

<>

Strategies to address risks

Following 11th September, new security strategies have emerged in many organisations which have been subject to these two sets of influences.[19] In some sectors of commerce the government has imposed compulsory security strategies by regulation, as discussed above. In other organisations the changing risks faced have made it prudent to introduce new security strategies. The example of Pan American Airways going out of business, which is largely attributed to terrorists blowing up flight 103 over Lockerbie in 1989,[20] and given that 60 to 90 per cent of organisations go out of business after experiencing a 'disaster',[21] illustrate the importance of developing the most effective strategies to deal with such an incident and to minimise the risks of it occurring.

It is important that an organisation has a clear policy setting out its plans for dealing with a major terrorist incident. The foundations for such a plan must be a risk assessment. Clearly the impact of 11th September on many organisations has impelled them to consider a much wider and more dangerous range of potential risks, ranging from 'rogue' civilian aircraft to CBNR attacks. Once this review has been undertaken, the task is to identify strategies to minimise the risk of such incidents occurring and to deal with them effectively if they do.

There are a number of situational crime prevention strategies that can be employed by organisations to reduce the opportunities for terrorist attacks. Clarke identified three types of such strategies to reduce the opportunities for criminals,[22] some of which also apply to terrorists. Primarily, there are measures to increase the effort of offending. Past terrorist attacks in the UK have led to 'target hardening' measures such as the 'ring of steel' that was implemented around the City of London following the Bishopsgate bombing, and on an individual level many have strengthened building structures, by for instance applying a protective film to glazing. 'Access control' measures such as the introduction of identity cards, automated access control systems or security officers to check passes have also been utilised in various locations.

At a second level there are measures to increase the risks to the offender of getting caught; given that 'new terrorism' employs suicidal terrorists this may not be as important, although it may influence a decision on whether an operation can successfully be mounted. Some of these strategies include 'entry/exit' screens where searches and screening are undertaken. The level of searches and screening have been increased in the aviation sector, as discussed above. There have also been some calls for greater El Al-style profiling of passengers before they can proceed 'airside'.[23] Further measures to increase the risks include numerous strategies to improve surveillance through CCTV, additional staff or changing the structure of the environment. The third level of crime prevention, reducing the rewards, is much more difficult to apply, particularly to 'new terrorism' where the terrorists' aim is not financial gain. The fundamental problem with situational measures, however, is displacement.[24] If 'new terrorists' are determined to carry out an attack, successful measures in one organisation, region or country may simply cause them to select a target elsewhere.

Increased security budgets

Underpinning the introduction of many of these strategies is the need for additional resources dedicated to security. In the immediate aftermath of 11th September one owner of a major tower in the USA was forced to spend $8 million on security in order to reassure tenants that it was safer than the twin towers. Tall buildings, high-profile locations and the aviation sector, along with many other ordinary businesses, have had to increase their security budgets.[25] However, one senior member of the British security industry has suggested that despite a huge interest in purchasing additional security immediately after 11th September across the full spectrum of organisations, it was only in the most vulnerable sectors such as aviation that there had been a significant increase.

This has not escaped the attention of many security companies seeking new markets. The poor state of aviation security in the USA has led some to speculate on increased opportunities for British manufacturers of equipment for detecting explosives, metal objects, etc.[26] Indeed, the British Security Industry Association (BSIA), with Trade Partners UK, has put together a brochure of British security solution providers for the American aviation industry.

Security and risk consultants have also experienced a huge demand for their services, particularly from large companies based in London.[27] Guidance has been sought on, *inter alia*, how an 11th September type of attack might affect the company, evacuation strategies in tall buildings, and disaster recovery procedures. Companies with staff and facilities abroad have also sought advice on the risks to their assets and staff.

The impact of 11th September has also led many individuals and organisations to invest in products they would not normally have considered purchasing. For instance, many have purchased gasmasks, fearing potential CBRN attacks. The private purchase of vaccines and drugs against certain biological attacks has also expanded.

Review of business continuity plans

A recent survey of 5000 companies found that only 45 per cent had business continuity plans (or 'consequence management' plans as they are also known) in place.[28] The minds of many organisations have been concentrated by 11th September, impelling them to review existing plans, or to develop a strategy if none had previously existed. For instance, largely as a result of experiencing Irish republican terrorism, many companies based in the City of London and Canary Wharf had plans, should their offices be destroyed, to relocate relatively quickly elsewhere in London. The attacks of 11th September, however, illustrated that the scale of a potential attack could be much greater, and consequently some companies have now arranged potential relocations elsewhere in the UK, and in some cases abroad. Similarly, the impact on Cantor Fitzgerald of 658 staff—virtually the whole department—being killed at the World Trade Centre has illustrated the need to imagine scenarios where an organisation loses a significant number of key staff. Some organisations have been training more staff to deal with such incidents, as well as conducting exercises.

Another relatively simple issue that the attacks of 11th September have highlighted is evacuation strategies, particularly in tall buildings. In the UK the experience of large PIRA lorry bombs exploding during the early 1990s has typically led to strategies to keep people in safe locations within buildings until the bomb has detonated. Clearly, after 11th September this would be a difficult measure to enforce. The attacks have also stretched the imagination to include a CBNR attack on a building. Such an attack would require different types of evacuation procedure, and many organisations have responded to this.

The attacks of 11th September also highlighted the way in which Western society is interconnected, interdependent and technologically dependent. They illustrated how small groups using asymmetric tactics could cause massive disruption leading to shock waves throughout the world. The attacks in New York and Washington DC had economic and psychological impacts far beyond the north-east seaboard of the USA.[29] They led to the evacuation of buildings and the grounding of aircraft well beyond the USA; in one incident, typical of many, a major company shut off its email system on 12th September because of the perception of a potential cyber-attack.

Although there has so far been no successful example of a major cyber-attack, there remains a risk of terrorist attacks on the communications infrastructure. In the UK, even before 11th September, the government had been taking steps to address this risk. The National Infrastructure Co-ordination Centre (NISCC) was established in 1999 to provide a single point of contact to co-ordinate the response to an electronic attack on the nation's critical national infrastructure. Its activities have been targeted, however, only at selected organisations. Additionally, in March 2002 the government appointed a Central Sponsor for Information Assurance and Resilience with a mandate to develop and implement a national strategy to ensure the security and resilience of information systems. The 11th September attacks, combined with government initiatives, have led some organisations to review their plans (or establish them where they did not already have them) for their key infrastructure should they be attacked. (Such a review by one organisation found that its main communications system actually shared the same network as its reserve—if one had been destroyed, so would the other.)

Insurance

The impact of 11th September on the insurance industry has been dramatic. The costs alone have been estimated by the US delegation to the OECD Insurance Committee at nearly $40 billion.[30] For the Lloyd's market in the UK, the net costs of have been estimated at £1.9 billion. The sheer scale of this has changed the conceptions of insurers on the magnitude of potential disasters, and thus on potential losses. The risk of such events will now have to take account of a potentially much larger number of casualties, and those organisations operating high-profile tall structures are likely to face greater responsibility in catering for potential hazards through disaster recovery plans. The consequences of all this for the policy-holder is that cover for terrorism is likely to be restricted, and perhaps in some cases not available at all, at a time when many organisations are likely to seek greater cover in such areas. The UK does have Pool Re, which is a re-insurance scheme of last resort, supported by the government, to provide cover for terrorist acts, though its scope was originally limited to property damage and consequential loss from fire and explosion. In July 2002 the government expanded the cover to biological contamination, impact by aircraft and flood damage, along with a number of other changes to support commerce.[31] Linked to changes in the scope of their policies, insurers are likely to examine the risk management strategies of organisations and to impose more conditions upon them, such as certain security measures. Therefore the legacy of 11th September has led to many organisations re-examining their insurance policies, and having to meet additional demands from insurers.

Review of staff travelling abroad

As the tragic events unfolded on 11th September, one major company attempted to track down all its staff working overseas. This proved extremely difficult, and as a consequence the company has now instituted a system whereby the locations of all staff overseas, and how they can be contacted, are known. The corporate security website of this company also provides up-to-date information on the risks, and is also linked to the Foreign and Commonwealth Office website providing information for foreign travellers. The potential risk of kidnapping in some regions of the world has also led some organisations to invest in more close-protection services.

Developing new partnerships

A tragic event such as 11th September places huge demands on the police, which they are not capable of dealing with alone. As Assistant Commissioner David Veness (of London's Metropolitan Police) summarised the situation when giving evidence to the House of Commons Defence Committee:

> If we need to evacuate a city, if we had a massive scene, or if, for example, there was a threat which required us to protect a sector of British industry which is pretty geographically spread, for example power, how would we go about that? We have not a gendarmerie. We have not got a third force … [or] a national guard.[32]

Such events would clearly require the help of the 'extended police family' such as special constables, police auxiliaries, private security staff, volunteers, etc. This would require strong partnerships, which cannot be created overnight but need to be developed over time. An enduring criticism of policing organisations—particularly in relation to the private security industry—is their lack of strong partnerships.[33] There have been some initiatives to improve this situation, but not enough. It is also likely that the proposed Civil Contingencies Bill will propose an obligation on local authorities and other agencies in the style of the Crime and Disorder Act—and probably also on the private sector—to co-operate in the preparation of contingency plans.

At another level the BSIA, which covers the majority of the British private security industry, has sought to develop a stronger partnership between its members in the sharing of intelligence. However, the decision makers in the security field need to develop stronger partnerships not only with the police but with other relevant government organisations. In this respect, though there is still much more to do, there are signs that this gap is beginning to be recognised; after 11th September, David Veness invited the heads of the ten leading guarding companies to establish a liaison council (these companies accounting for 15,000–20,000 security officers in London or its immediate vicinity).

Demands for greater access to intelligence

Linked to the issue of partnership is access to more intelligence for the wider security community. After 11th September, the Federal Bureau of Investigations (FBI) in the USA identified some of the key security decision makers, vetted them and brought them into a 'loop' where information and access to decisions would be provided. As Paul Wood, Head of Global Security at the third largest investment bank in Europe, UBS Warburg, told the House of Commons Defence Committee:

> [The British] government has continued to provide to the public sector advice and guidance on threat levels and how to protect its infrastructure and ... they have continued to provide that level of support and advice and guidance on threat levels [to the defence industry], but they really have not provided it to the private sector.[34]

On an individual basis Paul Wood had contacted the Cabinet Office, who had opened up a conduit for guidance and advice. Clearly this is an area where more open access needs to be provided—sector by sector, depending upon the level of threat—to sources of relevant information.

Tackling a culture of complacency

One of the biggest challenges that security decision makers have faced is tackling a culture of complacency. In many organisations there is a culture that however bad an attack might be it is 'unlikely to happen here'. In one organisation the senior security advisor described to the present authors the difficulty of getting some tenants (and personnel) to participate in evacuation drills. Clearly this poses one of the most significant challenges to any security decision maker: not only to keep the organisation's employees motivated to deal with a potential attack, but also to keep the attention of the board of directors so they dedicate the appropriate resources and put in place the right strategies to deal with the increased risk.

Conclusion

The events of 11th September have had dramatic economic, social and political consequences throughout the world. This chapter has reviewed the impact in terms of security-related issues on the UK business community. In providing this overview we have not provided an exhaustive list of the emerging strategies to combat this impact, rather the most significant. In doing so we hope to have provided some decision makers with ideas to improve or refine their response or to seek further information. We have also not considered some of the sectors of commerce that are not doing enough or even doing nothing. We also hope to have highlighted some areas where the government could make their task easier and more effective. Most notably, the foundation on which any sound security strategy must be based is an accurate risk assessment. If intelligence on the risks faced, amongst other relevant information, is not made available to security decision makers, how can they possibly implement effective security strategies? Greater access, subject to vetting and proper controls, should be provided. At another level, the introduction of higher standards for the private security industry, through the newly established Security Industry Authority, could further extend and improve the limited resources of the state to deal with a terrorist attack. Finally, although many organisations are taking the threat seriously and investing in new strategies, there are too many that are not. The government could also do more to encourage such organisations to take the threat more seriously; and in the most vulnerable sectors it should consider intervention. If there is any certainty in the events of 11th September it is that the terrorists will attempt to strike again. As it is the business organisations that are most likely to be the target in some form, it is essential that we seek to maximise the effectiveness of their response—and much more can be done to achieve this.

Notes

1 Rt Hon Bruce George is the Member of Parliament for Walsall South and Chairman of the House of Commons Defence Committee. Mark Button is Senior Lecturer in the Institute of Criminal Justice Studies, University of Portsmouth. Natalie Whatford is a Research Assistant to Bruce George; email: georgeb@parliament.uk.

2 House of Commons Defence Committee (2001) *The Threat from Terrorism.* HC 348-I. Vol. I. London: Stationery Office.

3 See Wilkinson, P. (2001) *Terrorism Versus Democracy.* London: Frank Cass.

4 House of Commons Defence Committee, op cit, p xvii.

5 Wilkinson, op cit.

6 Beck, A. and Willis, A. (1994) The Changing Face of Terrorism: Implications for the Retail Sector. In Gill, M. (ed.) *Crime at Work: Studies in Security and Crime Prevention.* Leicester: Perpetuity Press.

7 See House of Commons Defence Committee (2002) Defence and Security in the UK. HC 518-I. Vol. 1. London: Stationery Office, p 16.

8 Ibid, pp 40–54.

9 Wilkinson, op cit, p 208.

10 US Department of State (2002) *Patterns of Global Terrorism 2001.* Washington, DC: US Department of State.

11 Helminger, P. (2002) *The Economic Consequences of 11 September and the Economic Dimension of Terrorism.* Brussels: NATO Parliamentary Assembly.

12 House of Commons Defence Committee (2002) op cit, p 36.

13 Leppard, D. and Fielding, N. (2002) British Link to Muslim Terror Ranches in US. *Sunday Times*, 28th July.

14 House of Commons Defence Committee (2002) op cit, p 39.

15 Ibid.

16 Button, M. and George, B. (2001) Government Regulation in the United Kingdom Private Security Industry: The Myth of Non-Regulation. *Security Journal.* Vol. 14, No. 1, pp 55–66.

17 House of Commons Defence Committee (2002) op cit.

18 George, B. and Button, M. (2000) *Private Security.* Leicester: Perpetuity Press, p 128.

19 Helminger, op cit.

20 Broder, J.F. (2000) *Risk Analysis and the Security Survey.* Boston, MA: Butterworth Heinemann.

21 Hiles, A. and Barnes, P. (2001) *The Definitive Book of Business Continuity Management.* Chichester: Wiley.

22 Clarke, R.V.G. (ed.) (1992) *Situational Crime Prevention: Theory and Practice.* New York: Harrow and Heston.

23 El Al security staff ask passengers a series of questions to identify potential terrorists. It has been claimed the alleged 'shoe bomber' Richard Reid would never have been allowed to board the plane in Paris if it had belonged to El Al.

24 Reppetto, T. (1976) Crime Prevention and the Displacement Phenomenon. *Crime and Delinquency*. Vol. 22, No. 1, pp 166–77.

25 Helminger, op cit.

26 Mackay, D. and Mowbray, C. (2002) How Legacy of September 11 may Benefit Britain. *Public Security*. No. 16, Spring, pp 14–15 .

27 Unpublished evidence supplied by the Association of Security Consultants to the House of Commons Defence Committee.

28 House of Commons Defence Committee (2002) op cit, p 39.

29 Helminger, op cit.

30 Ibid.

31 Association of British Insurers (2002) News Release, 23rd July.

32 House of Commons Defence Committee (2002) op cit, p 57.

33 Button, M. (2002) *Private Policing*. Cullompton: Willan; Stenning, P.C. (1989) Private Police and Public Police: Toward a Redefinition of the Police Role. In Loree, D.J. (ed.) *Future Issues in Policing*. Symposium Proceedings. Ottawa: Canadian Police College.

34 House of Commons Defence Committee (2002) op cit, p 29.

Index

('f' denotes a figure and 't' a table)